THE CONTRAST MEDIA MANUAL

THE CONTRAST MEDIA MANUAL

Edited by
RICHARD W. KATZBERG, M.D.
Professor and Chair
Department of Radiology
University of California, Davis, Medical Center
Sacramento, California

Williams & Wilkins
BALTIMORE • PHILADELPHIA • HONG KONG
LONDON • MUNICH • SYDNEY • TOKYO
A WAVERLY COMPANY

Editor: Timothy H. Grayson
Managing Editor: Marjorie Kidd Keating
Copy Editor: Bill Cady
Designer: Norman W. Och
Illustration Planner: Ray Lowman
Production Coordinator: Raymond E. Reter

Copyright © 1992
Williams & Wilkins
428 East Preston Street
Baltimore, Maryland 21202, USA

All rights reserved. This book is protected by copyright. No part of this book may be reproduced in any form or by any means, including photocopying, or utilized by any information storage and retrieval system without written permission from the copyright owner.

Accurate indications, adverse reactions, and dosage schedules for drugs are provided in this book, but it is possible that they may change. The reader is urged to review the package information data of the manufacturers of the medications mentioned.

Printed in the United States of America

Library of Congress Cataloging-in-Publication Data

The Contrast media manual / edited by Richard W. Katzberg.
    p. cm.
  Includes bibliographical references and index.
  ISBN 0-683-04542-3
  1. Contrast media—Handbooks, manuals, etc. 2. Contrast media—Diagnostic use—Side effects—Handbooks, manuals, etc.
I. Katzberg, Richard W.
  [DNLM: 1. Contrast Media—handbooks. 2. Diagnostic Imaging—methods—handbooks. WN 39 C764]
RC78.7.C65C67 1992
616.07′54—dc20
DNLM/DLC
for Library of Congress   91-21624
                                                                 CIP

                                                                       94   95
                                      2   3   4   5   6   7   8   9   10

Dedicated

with

love and appreciation

to

my mother and father

Mary Frances Wier Katzberg, R.N.

and

Arthur Joseph Katzberg, M.D.

Clinton, South Carolina

This Edition has been translated into Japanese by:
    Keiko Sakuyama, MD
    Hitoski Katayama, MD
    Hirotsugu Munechika, MD

PREFACE

The objective of this small text is to provide a practical, simple, "nuts-and-bolts" guide for daily clinical use of contrast media from recognized authorities in the field.

There were two stimuli for organizing this material. First, I had the unusual opportunity of enjoying a 12-year experience as both resident and faculty member in a department that was seriously involved in contrast media research and practical clinical applications. This department had a substantial number of existing and visiting professors who were expert in the field. I believed there would be some usefulness in translating the many theories and clinical experiences commonly discussed in this unique environment into a practical approach to daily patient care.

The second stimulus occurred by circumstance. While attending the American Roentgen Ray Society meeting in New Orleans in May of 1989, I ran into a bright and enthusiastic young editor, Timothy Grayson, who shared the same practical philosophy as I. Over a fried catfish sandwich lunch at Fitzgerald's, we agreed on the project.

The reader will note some redundancy in the material between chapters, e.g., tables of contrast media. This has been maintained for the convenience of the reader and to provide chapters that are self-contained and philosophically specific for the contributors.

I would like to acknowledge the great expertise of the authors whose contributions are derived as a natural consequence of each individual's extensive clinical experience. In addition, I would like to offer my thanks to the many expert radiologists dedicated to contrast media research whom I have had the opportunity of meeting and who have provided excellent role models. From the University of Rochester School of Medicine and Dentistry, Rochester, New York, these people are: Drs. Harry Fischer, Tom Morris, Beverly Wood, Francis Burgener, Zoran Barbaric, Joe Skucas, Phil Harnish, Martti Kormano, Peter Dean, and George Foote. From the Brigham and Women's Hospital, these people are: Drs. Bill Caldicott, Herb Abrams, Norm Hollenberg, and Harry Mellins. Other notable colleagues from other academic institutions include: Drs. Elliott Lasser, Lee Talner, Milos Sovac, Tony Lalli, Geoff Benness, and Tom Sherwood.

Many thanks to the dedicated staff at Williams & Wilkins, Baltimore, Maryland, and for the excellent secretarial assistance of Marilyn Aberle and Lori Neville.

CONTRIBUTORS

Albert Alexander, M.D.
Clinical Assistant of Radiology
Massachusetts General Hospital
Clinical Fellow
Harvard Medical School
Boston, Massachusetts

Michael A. Bettmann, M.D.
Professor of Radiology
Chief, Cardiovascular Imaging
Department of Radiology
Boston University Medical Center/
  University Hospital
Boston, Massachusetts

William H. Bush, M.D.
Professor and Director
Section of Genital Urinary Radiology
Department of Radiology
University of Washington School of
  Medicine
Seattle, Washington

N. Reed Dunnick, M.D.
Professor of Radiology
Director, Division of Diagnostic
  Imaging
Department of Radiology
Duke University Medical Center
Durham, North Carolina

Sven Ekholm, M.D., Ph.D
Department of Radiology
University of Gothenburg
Sahlgrenska Hospital
Gothenburg, Sweden

Harry W. Fischer, M.D.
Professor Emeritus
Department of Radiology
University of Rochester Medical
  Center
Rochester, New York

Edmund A. Franken, Jr., M.D.
Professor and Head
Department of Radiology
The University of Iowa College of
  Medicine
Iowa City, Iowa

Robert R. Hattery, M.D.
Consultant in Diagnostic Radiology
Professor, Mayo Medical School
Mayo Clinic and Mayo Foundation
Rochester, Minnesota

Peter D. Jacobson, J.D., M.P.H.
Rand Corporation
Santa Monica, California

Richard W. Katzberg, M.D.
Professor and Chair
Department of Radiology
University of California, Davis,
  Medical Center
Sacramento, California

John A. Kaufman, M.D.
Fellow, Vascular/Interventional
  Radiology
Department of Radiology
Boston University Medical Center
Boston, Massachusetts

Bernard F. King, M.D.
Senior Associate Consultant in
  Diagnostic Radiology
Assistant Professor, Mayo Medical
  School
Mayo Clinic and Mayo Foundation
Rochester, Minnesota

Anthony F. Lalli, M.D.
Niagara-on-the-Lake
Ontario, Canada

Contributors

Elliott C. Lasser, M.D.
Professor
Department of Radiology
University of California, San Diego
La Jolla, California

Richard A. Leder, M.D.
Assistant Professor
Department of Radiology
Duke University Medical Center
Durham, North Carolina

Thomas W. Morris, Ph.D.
Associate Professor of Radiology and Physiology
Research Director for Radiology
Department of Radiology
University of Rochester Medical Center
Rochester, New York

C. John Rosenquist, M.D.
Professor
Department of Radiology
University of California, Davis, Medical Center
Sacramento, California

Daniel I. Rosenthal, M.D.
Director, Bone and Joint Radiology
Department of Radiology
Massachusetts General Hospital
Associate Professor of Radiology
Harvard Medical School
Boston, Massachusetts

Val M. Runge, M.D.
Rosenbaum Professor of Diagnostic Radiology
Director, Magnetic Resonance Imaging and Spectroscopy Center
University of Kentucky Medical Center
Lexington, Kentucky

Arthur J. Segal, M.D.
Associate Chairman
Department of Diagnostic Radiology and Nuclear Imaging
Rochester General Hospital
Clinical Associate Professor of Radiology and Urology
University of Rochester School of Medicine and Dentistry
Rochester, New York

Jovitas Skucas, M.D.
Professor
Department of Radiology
University of Rochester Medical Center
Rochester, New York

Wilbur L. Smith, Jr., M.D.
Professor of Radiology and Pediatrics
Vice Chairman, Director of Clinical Services
Department of Radiology
University of Iowa College of Medicine
Iowa City, Iowa

Amy S. Thurmond, M.D.
Assistant Professor of Radiology and Obstetrics and Gynecology
Director of Women's Imaging
Director of Ultrasound
Oregon Health Sciences University
Portland, Oregon

CONTENTS

Preface... vii
Contributors... ix

Chapter / 1
Intravascular Contrast Media: Properties and General Effects .... / 1
Thomas W. Morris and Richard W. Katzberg

Chapter / 2
Treatment of Acute Reactions to Contrast Media ...... / 19
William H. Bush

Chapter / 3
Contrast Medium-Induced Nephrotoxicity .... / 28
Richard W. Katzberg

Chapter / 4
Imaging in the Patient with Azotemia .... / 36
Arthur J. Segal

Chapter / 5
Urography, Cystography, and Urethrography .... / 50
Bernard F. King and Robert R. Hattery

Chapter / 6
Contrast Media Use in Computed Tomography ...... / 66
Richard A. Leder and N. Reed Dunnick

Chapter / 7
Vascular Contrast Media Use in the Central Nervous System .... / 83
Sven Ekholm

Chapter / 8
General Angiography . . . . / 94
   Michael A. Bettmann

Chapter / 9
Peripheral Venography . . . . / 102
   John A. Kaufman and Michael A. Bettmann

Chapter / 10
Myelography . . . . / 107
   Sven Ekholm

Chapter / 11
Arthrography . . . . / 117
   Albert Alexander and Daniel I. Rosenthal

Chapter / 12
Hysterosalpingography . . . . / 131
   Amy S. Thurmond

Chapter / 13
Magnetic Resonance Contrast Agents . . . . / 143
   Val M. Runge

Chapter / 14
Mechanisms of Contrast Media Reactions: Implications for Avoidance and Treatment Based on Hypothesis of Causation . . . . / 161
   Harry W. Fischer (Part I)
   Anthony F. Lalli (Part II)
   Elliott C. Lasser (Part III)

Chapter / 15
Low Osmolar Contrast Agents: Economic and Legal Issues . . . . / 180
   C. John Rosenquist and Peter D. Jacobson

Chapter / 16
Barium Sulfate for Gastrointestinal Use . . . . / 187
   Jovitas Skucas

Chapter / 17
Water-Soluble Gastrointestinal Contrast Agents . . . . / 200
   Jovitas Skucas

Chapter / 18
Gastrointestinal Agents in Computed Tomography . . . / 202
    Jovitas Skucas

Chapter / 19
Fistulographic Contrast Agents . . . . / 205
    Jovitas Skucas

Chapter / 20
Pediatric Contrast Agents . . . . / 207
    Wilbur L. Smith, Jr., and Edmund A. Franken, Jr.

Index . . . . / 223

Chapter / 1
Intravascular Contrast Media: Properties and General Effects

Thomas W. Morris
Richard W. Katzberg

I. **General considerations.** Intravascular contrast media are administered to more than 6 million patients in the United States on a yearly basis in a worldwide annual volume of several thousand metric tons, easily outweighing any other pharmaceutical. The indications have continued to increase: examinations requiring contrast media include excretory urography, head and body computed tomography (CT), angiography, digital subtraction angiography (DSA), and magnetic resonance imaging (MRI). The basic aspects of iodinated intravascular contrast media are presented here, and the basic aspects of intravascular MRI agents are discussed in Chapter 13.

The use of iodinated water-soluble agents is not based on their pharmacologic action but on their distribution in and elimination from the body. Therapeutic agents are given in very small quantities at regularly spaced intervals. Contrast media are used in quantities as large as 100 gm and are most often administered as a bolus lasting only 1 or 2 minutes. Therapeutic drugs are given to effect biologic or chemical change, whereas contrast media are used for organ or tissue enhancement. Most intravascular drugs are nearly isotonic, yet contrast media have osmolalities up to 7 times that of body fluids (osmolality of body fluids is roughly 290 mOsm/kg of water, whereas the osmolality of contrast media ranges from 600 to 2200 mOsm/kg of water). This difference can produce osmotic pressure gradients measured in thousands of mm Hg (*Radiographic Contrast Agents.* Gaithersburg, MD: Aspen Publishers, 1989, pp. 119–128). **High osmolality** and **viscosity** are the major characteristics of water-soluble contrast media that are responsible for the hemodynamic, cardiac, and subjective effects. Sudden drastic water shifts from the interstitial and cellular spaces into the plasma are a result of osmotic pressure gradients and explain many of the adverse effects of contrast media. **These effects include vasodilatation, heat, pain, a variety of hemodynamic effects, and an osmotic diuresis.**

Although twice the osmolality of blood, the newer contrast agents are of relatively low osmolality (have fewer particles in solution) compared with the older agents. Thus, new agents have been dubbed **"low osmolar contrast media" (LOCM),** whereas older agents are called **"high osmolar contrast media" (HOCM).**

II. **Physical characteristics.** All radiographic contrast agents are clear, **colorless** fluids with no precipitate in the contrast media vial. The commonly used term **"dye"** is the layman's terminology for the contrast medium but is a misnomer.

The **viscosity** or "friction" of the medium is an important physical property that influences the injectability or delivery through small-bore needles and angiographic catheters. Viscosity is influenced by the concentration and basic chemical design of the molecule. Sodium salts are less viscous than meglumine salts. Heating the contrast medium to body temperature just prior to injection significantly decreases the viscosity and facilitates the capability of rapid injection.

III. **Physical principles of contrast media related to imaging**
   A. **Radiographic contrast media** attenuate a beam of X-ray radiation according to the relationship illustrated in Figure 1.1. **The linear attenuation coefficient in this equation depends on the elements involved, their concentration, the thickness of the material, and the energy (keV) of the X-ray radiation.** The iodine atoms in the contrast medium molecule are the primary attenuators of the X-ray radiation, and the linear attenuation coefficient is proportional to the mg of iodine per mL of solution. The attenuation of equal thicknesses of dilute contrast media and water is shown as a function of keV in Figure 1.2. The "step" change in the iodine curve at 33.2 keV is related to the K-shell-binding energy (termed **"K edge"**) for iodine (*Introduction to the Physics of Diagnostic Radiology.* Philadelphia: Lea & Febiger, 1984, pp. 60–76). In clinical radiographic examinations, iodine provides greater attenuation relative to bone or tissue between 33 and 60 keV than at higher keV. Contrast media attenuate the X-ray beam and thus decrease the amount of X-ray radiation reaching the X-ray detector. The X-ray detector can be film, a film/screen combination, an image intensifier,

INTRAVASCULAR CONTRAST MEDIA: PROPERTIES AND EFFECTS 3

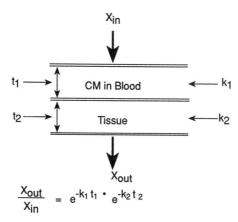

Figure 1.1. As the X-ray beam passes through body structures, it is attenuated as illustrated for a monoenergetic X-ray source and a two-component structure (contrast media (CM) in blood and tissue) with thicknesses t_1 and t_2 and linear attenuation coefficients k_1 and k_2. The linear attenuation coefficient is a nonlinear function of X-ray energy (keV) as shown in Figure 1.2. Since the output of an X-ray tube contains a broad band of energies and the body is composed of many elements with widely different attenuation coefficients, the theoretical prediction of the attenuation (x_{out}/x_{in}) is complex.

or a xenon detector in a CT system. CT systems have the most sensitive detectors, while film is the least sensitive.

V. Contrast media molecules and additives. The most widely used angiographic and urographic contrast agents are water-soluble, fully substituted, triiodinated benzoic acid derivatives. Intravascular contrast media molecules are commonly classified according to two characteristics: (1) **ionic** versus **nonionic** and (2) **monomer** versus **dimer.** Ionic contrast media are salts similar to sodium chloride that form ions when they are in water solutions. Currently available ionic contrast media dissociate into monovalent anions and cations. Nonionic contrast media do not have anion and cation components and therefore do not dissociate or ionize in water. Figure 1.3 illustrates the four general types of contrast media molecules available today. These include **monoacidic monomer (ionic monomer), nonionic monomer, monoacidic dimer (ionic dimer), and nonionic dimer.** The differences between contrast media molecules within a single type are related to the organic side chains. The ratio of iodine atoms to dissolved particles is an important characteristic of contrast media and is a commonly used terminology in the

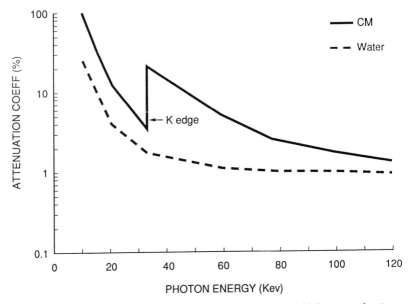

Figure 1.2. Relative linear attenuation coefficients for equal thicknesses of water and dilute contrast media (*CM*) are given as a function of photon energy (keV). The keV dependence of iodine provides a relatively good match to the photon energy distribution of an X-ray tube operating at 60–100 kVp.

literature. A higher ratio is more desirable, since more iodine means better opacification and fewer particles mean lower osmolality. Newer contrast media have a desirable 3:1 ratio, whereas traditional high osmolar agents have a lower ratio of only 3:2 (or 1.5:1). A logical, new generation of LOCM is formulated with the ratio of 6:1. These agents can be manufactured as isotonic solutions.

Commercially available contrast media formulations also contain trace amounts of **free iodine, heavy-metal scavengers** such as **calcium disodium-EDTA,** and residuals of the solvents **ethylene glycol** and **DMSO.** There is also an addition of electrolytes to the formulation for pH control and other purposes.

All of the contrast media molecules currently in clinical use are among the safest pharmaceutical substances ever synthesized; however, these compounds are also used in relatively massive doses. This helps to explain how even relatively harmless compounds can have significant toxicity.

V. **Physical and chemical properties of contrast media solutions.**
The physical and chemical properties of contrast media solutions depend on their concentration. Package inserts for X-ray

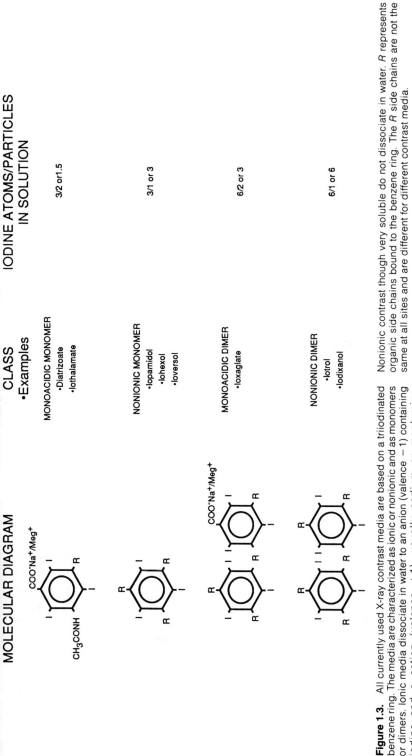

Figure 1.3. All currently used X-ray contrast media are based on a triiodinated benzene ring. The media are characterized as ionic or nonionic and as monomers or dimers. Ionic media dissociate in water to an anion (valence −1) containing iodine and a cation (valence +1), usually sodium or meglumine. Nonionic contrast though very soluble do not dissociate in water. R represents organic side chains bound to the benzene ring. The R side chains are not the same at all sites and are different for different contrast media.

contrast media generally give concentration as mg of iodine per mL of solution. This specification is only correct at the stated temperature. As the temperature increases, the volume of the solution will increase, and the mg of the iodine per mL will decrease; however, this effect is small and of no real clinical significance. The most familiar chemical measure of concentration is moles of solute per liter of solution (molar or M). The **molar concentration** of a contrast media solution can be calculated by dividing the mg of iodine per mL by the number of iodine atoms per molecule times the molecular weight of iodine (approximately equal to 127). Molar concentrations also decrease as solution temperature increases. A measure of concentration independent of temperature is moles of solute per kg of water **(molal)**. A measure of concentration that is of great biologic significance is **osmolality**, which is the total number of particles in solution per kg of water. Iodine concentrations given on the package insert of the contrast media vial can be used to calculate the approximate **osmolality** of the various agents. As an example, for **monomeric** agents,

$$\text{Osmolality (mOsm/kg)} = \left(\frac{\text{mg I/mL}}{N \times 127}\right) \times K \times G$$

Where N is the number of iodine atoms per molecule, 127 is the molecular weight of iodine, K is the number of particles in solution (2 for ionic and 1 for nonionic media), and G is the osmotic coefficient that varies with concentration and with specific media (for most media, G is between 0.80 and 1.0). Thus the newer, nonionic media have lower tonicity for a given concentration of iodine, since K is half as large. Dimers have 6 iodine atoms per molecule, compared with 3 for monomers, so the osmolality of dimers is half of that for monomers. The osmolality of a solution determines osmotic pressure, which controls water movement.

A. Density. The density of a solution is the weight divided by volume; i.e., gm/mL. The linear attenuation coefficients for contrast media solutions are directly proportional to the density of these solutions. Contrast media solution density is dominated by the iodine atoms, and therefore, the density of all solutions is fairly similar at the same content of iodine per volume of solution.

B. Viscosity. The viscosity of a solution is a measure of the solution's resistance to deformation in response to shear forces, such as those created by pressure differences. Viscosity is a nonlinear function of solution molality. As molality increases, the viscosity tends to increase faster than linear. The viscosity of a contrast media solution decreases rapidly when temperature is increased from room to body temperature.

C. **Osmolality.** The osmolality of a contrast media solution is equal to the number of particles per kg of water. Under ideal conditions, the osmotic forces causing water movement are directly proportional to differences in osmolality. The osmolality of a solution under practical conditions is not ideal and thus must be measured. In part, this is because single molecules may form weak associations and ionic molecules do not completely dissociate in water. Many papers erroneously use the term **osmolarity** to describe contrast media osmotic concentrations. Osmolarity is a calculated value of the ideal number of particles per liter of water. Table 1.1 lists properties of some currently used intravascular contrast media.

D. **Partition coefficient.** The lipid solubility of molecules effects their distribution and biologic toxicity. The lipid solubility is usually described as the distribution "or partition" of the molecules in a mixture of n-butanol and water. The partition coefficient is the amount of the molecule in the butanol fraction, divided by the amount in the water fraction.

Table 1.1. Properties of Representative Contrast Media

Molecule	Diatrizoate	Ioxaglate	Iohexol	Iotrol
Type	Ionic monomer	Ionic dimer	Nonionic monomer	Nonionic dimer
Iodine atoms per particles in solution	1.5	3	3	6
Molecular weight	613	1209	821	1626
Partition coefficient[a]	0.045	0.104	0.070	0.005
Density at 20°C for 300 mg I per mL (gm/mL)	1.34	1.32	1.35	1.35
Viscosity at 37°C for 300 mg I per mL (mPa·sec)[b]	4.2	6.2	6.3	9.1
Osmolality at 300 mg I/mL (Osm/kg)	1.57	0.56	0.67	0.36
Acute LD_{50} in mice (gm/kg)[c]	7.5	13.4	24.2	26.0

[a]Lipid solubility is measured by the partition of the molecule in a mixture of n-butanol and water.
[b]Water at room temperature has a viscosity of 1 cp or 1 mPa·sec.
[c]The dose of contrast media producing acute death in 50% of mice, given via a 1-minute IV injection.

E. **Acute lethal dose (LD_{50}).** The acute LD_{50} is the dose of a contrast medium required to cause a mortality rate of 50% following an intravenous injection of 1 minute duration. Mortality is usually observed for 24 hours for contrast media testing. The LD_{50} for contrast media is usually expressed as the gm iodine per kg body weight. The injection duration is an important variable. At equal doses, mortality increases as duration decreases.

VI. **Pharmacokinetics.** Contrast media, because of their molecular size and extremely low chemical reactivity with body fluids and tissues, have pharmocokinetics very similar to a class of compounds termed **"extracellular tracers."**

  A. **Distribution volume.** All of the current X-ray contrast media molecules have very low lipid solubility and very low chemical reactivity and range in molecular weight from 600 to 1650. Molecules of this type distribute throughout the body's **extracellular space.** There is no available evidence of any significant amount of penetration of these contrast media molecules through the cell membrane into the interior of viable cells. Theoretically, the differences in the molecular weight and size between the monomer and dimer contrast media could cause slightly slower distribution in the extracellular space for the larger dimers. This effect, however, appears to be quite small and is not of any clinical significance, even if measurable. Following the intravascular injection of contrast media, the plasma concentration of iodine follows a **biexponential decay curve** (Fig. 1.4). The first exponential term describes the mixing of the contrast media in the plasma volume and then its distribution into the interstitial space. The second exponential term represents the clearance of the contrast medium molecules from the body.

  B. **Clearance.** The clearance of X-ray contrast media and other extracellular tracers is primarily by **glomerular filtration** and **renal clearance.** None of the molecules are reabsorbed or secreted by the renal tubules (Fig. 1.5). In cases of complete cessation of renal function, elimination of the contrast medium is through the liver and gut (**vicarious excretion**), which occurs at a much slower rate than via the kidneys. Under normal physiologic conditions, very close to 100% of the contrast medium is eliminated through the kidney, and the instantaneous rate of removal is equal to the glomerular filtration rate times the plasma iodine concentration. The clearance of contrast media molecules is usually described by

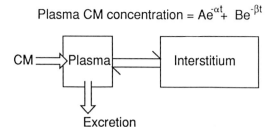

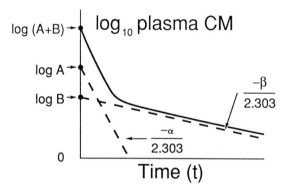

$$\text{Excretion half time} = \frac{0.693}{\beta}$$

$$\text{Total distribution volume} = \frac{\text{dose(amount/kg)}}{B \text{ (amount/L)}}$$

$$\text{Plasma volume} = \frac{\text{dose}}{A + B}$$

Figure 1.4. Intravascular contrast media (*CM*) are injected into the plasma space, and the concentration equilibrates between the plasma and the interstitium. Excretion occurs via the kidney and glomerular filtration. The plasma concentration time curve is approximated by a biexponential equation. The total distribution volume approximates the extracellular space. *A* and *B* are the time zero intercepts of the exponential components.

the **half-time** for the renal clearance portion of plasma decay curves. The **half-time** in patients with normal renal function is between 1 and 2 hours for all of the classes of contrast agents. In the absence of renal function the extracellular concentration of iodine will approach a value given by the

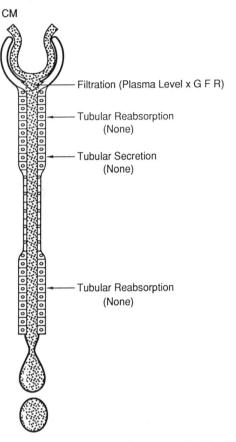

Figure 1.5. Intravascular contrast media (*CM*) are handled by the kidney in the same way as inulin. Contrast medium filters through the glomerulus into the tubules, where it is not reabsorbed or secreted. However, water, electrolytes, and other solutes may be reabsorbed.

total iodine injected, divided by the extracellular volume, which is approximately 200 mL/kg body weight.

VII. Intravenous contrast media injection techniques

A. Injection rates and plasma contrast media concentration. Intravenous injections of contrast media are generally characterized as either bolus injections or infusions. The terms are used rather loosely, and the important fact is that during rapid injections of contrast media into the venous system, the plasma iodine concentration reaches a much higher peak, although it is very transient or decays rapidly. During prolonged injections that we call infusions, aortic plasma iodine is much lower but more constant (Fig. 1.6). The transit of the contrast media through the pulmonary system greatly influences aortic iodine concentrations. If the injection durations are shorter than this transit time, which is generally about 10 seconds, the bolus will complete transit of the pulmonary system before recirculation occurs. If the injection is of longer duration than the transit time through the lungs, recirculated iodine will add to the iodine being injected. During infusions, there is thus a constant increase in the aortic plasma iodine concentration throughout the injection or until the rate of excretion equals the rate of injection. Bolus injections of contrast media in the intravenous system are very useful for intravenous DSA and for dynamic CT. Experimental studies have shown that for maximum aortic iodine the duration of injection should be in the range of approximately 2 seconds. The final aortic iodine

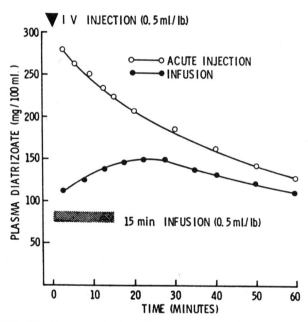

Figure 1.6. The plasma contrast medium concentration-time curve depends on the rate of injection. A rapid bolus (< 1 minute) produces a high initial peak and a rapid decline. A slow 15-minute infusion produces a continual increase during the injection but never reaches the high levels seen with a bolus. (From W. R. Cattell. Excretory Pathways for Contrast Media. *Investigative Radiology* 5(6):473–497, 1970.)

concentration is going to be a function of both cardiac output, central blood volume, and the amount of iodine injected (*Investigative Radiology* 18:308–316, 1983).

B. Physiologic actions of intravenous contrast media

1. During a rapid bolus injection of contrast media into the venous system, there is a rapid increase in the osmolality of the plasma because of the hypertonic contrast media. This increase in plasma osmolality causes a movement of water from red blood cells (RBCs) and from pulmonary tissue into the plasma space in response to a very large osmotic gradient. These fluid shifts cause a drop in hematocrit and an increase in plasma volume. After the hypertonic mixture of blood and contrast media transits the pulmonary vascular system, there is a reversal of the osmotic pressure gradient as blood without contrast media flows into the pulmonary tissue. Thus, water moves back into the pulmonary tissue. If the initial osmotic stress causes significant endothelial cell shrinkage in the pulmonary capillaries, the intercellular clefts may open and allow proteins to leak into the pulmonary interstitial space. These conditions would cause an additional accumulation of fluid in the lungs.

2. When the contrast media are entering the pulmonary vasculature and immediately after, there is generally an elevation in pulmonary arterial pressure (PAP). After the contrast media reach the systemic circulation, there is an increase in cardiac output and a decrease in peripheral and pulmonary vascular resistances. Systemic arterial pressure may decrease some, but the amount is quite variable. All of these transient cardiovascular changes last only a few minutes and are of greater magnitude with more hypertonic contrast media (*Radiology* 149:371–374, 1983).

3. **When the contrast media reach the systemic microcirculation, the molecules quickly equilibrate across capillary membranes (except an intact blood-brain barrier) (BBB) (Fig. 1.7).** In the first phase of distribution, the increase in intravascular osmolality causes a rapid fluid shift across capillary membranes toward the hypertonic (intravascular) compartment. Since the agents cannot cross normal cell membranes, cellular dessication or "dehydration," particularly of RBCs, is one possible effect of this shift. A net loss of intracellular water may be one mechanism

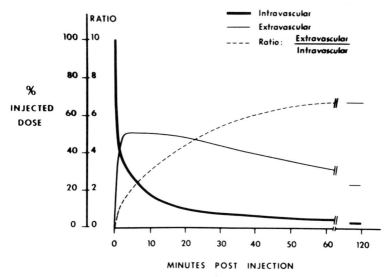

Figure 1.7. There is a rapid redistribution of the contrast media from the plasma to the interstitial space. The data presented are for the distribution of iothalamate in the dog. (From J. H. Newhouse. Fluid Compartment Distribution of Intravenous Iothalamate in the Dog. *Investigative Radiology* 12:364–367, 1977.)

responsible for the generalized stimulation of peripheral receptors, which results in the sensation of heat, pain, and involuntary movement patients so often experience. The rapid influx of water from the RBCs, the interstitial space, and parenchymal cells into the vascular compartment causes a rapid increase in blood volume of up to 16%, a decrease in hematocrit, an increase in cardiac output, and a decrease in peripheral vascular resistance (Fig. 1.8).
4. The changes in hemodynamics are dependent on the contrast media osmolality, the rate of injection, and the type of contrast agent. Ionic contrast agents have been shown to bind ionic calcium, and it is possible that this may cause some additional transient myocardial depression during bolus IV injections.

II. Intra-arterial injections
A. Injection rate and blood-contrast media mixing.
Intra-arterial injections of contrast media cannot take advantage of blood-contrast media mixing in cardiac chambers. During intra-arterial injections, adequate mixing of the contrast media and blood requires that the injection rate be close to the normal flow rate of blood in the artery. The final iodine

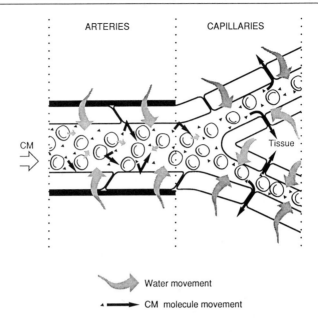

Figure 1.8. When the hypertonic contrast medium (*CM*) is injected into blood, there is a rapid movement of water from red blood cells, endothelial cells, and tissue into the plasma. As the bolus of contrast medium enters capillaries, the movement of water continues, and the contrast medium molecules move through the clefts between endothelial cells into the interstitium. Hypertonicity seems to be the major factor producing a number of physiologic changes.

concentration in the artery is approximately equal to the injection rate divided by the blood flow rate times the contrast medium iodine (*Investigative Radiology* 24:361–365, 1989). When the injection rate is greater than the blood flow rate, the iodine concentration in the artery is equal to the concentration of the contrast media injected.

B. Physiologic effects of intra-arterial injections
   1. The contrast media injection jet through the catheter slightly increases distal pressure and flow but greatly

reduces proximal flow into the artery, since the contrast media injected replace the blood flow.
2. Although the viscosity of blood in large arteries is about 4 cp, as the blood reaches the arterioles, the viscosity approaches that of plasma (approximately equal to 1.2 cp). All contrast media have higher viscosities than that of plasma; thus the contrast media viscosity generally causes an increase in resistance and a decrease in flow during the period in which the contrast media pass through the arterioles (*Investigative Radiology* 17:70–76, 1982).
3. The high osmolality of the contrast media-blood mixture flowing in the artery causes movement of water from RBCs to plasma and from tissue to plasma. The contrast media molecules filter into the interstitial fluid through the endothelial cell clefts (Fig. 1.8). In the cerebral circulation the so-called BBB greatly limits the movement of contrast media molecules into the brain interstitial space. After the contrast media-blood mixture transits the capillaries in the vascular bed, the incoming blood is of lower osmolality, the fluid flux reverses to the tissue, and contrast media molecules reverse their movement from the interstitial space to the plasma (*Investigative Radiology* 18:335–340, 1983).
4. Intracoronary injections. Intracoronary injections of ionic contrast media cause a decrease in heart rate, a decrease in peak left ventricular (LV) pressure, a decrease in the maximum rate of increase in LV pressure, an increase in LV end-diastolic pressure, an increase in LV end-diastolic diameter, a decrease in systemic arterial pressure, and a biphasic decrease then increase in coronary blood flow. Injection of ionic contrast media into the coronary artery causes a direct inhibitory effect on the sinoatrial node. A similar inhibitory effect has been observed in conduction through the bundle of His and Purkinje fibers. To a lesser extent, the decline in heart rate during coronary arteriography is cholinergically mediated. Nonionic contrast media do not cause the large myocardial depression seen with ionic media and may actually cause increased contractility. Nonionic contrast media, since they have very low electrolyte concentrations, also have different effects on electrophysiology. However, both nonionic and

ionic media can cause significant electrical disturbances.
5. Injections in carotid and vertebral arteries. Injections of hypertonic contrast media into the carotid and vertebral arteries can cause a decrease in the pulse rate, periods of asystole, and lowering of the arterial blood pressure. For these procedures it has been shown that the meglumine salts of ionic contrast media have fewer cardiovascular effects than sodium salts; however, nonionic media are considered even safer.

IX. **Contrast media penetration into the central nervous system (CNS).** The CNS is separated from substances in the blood by means of the blood-brain barrier (BBB). The endothelial cells of the CNS capillary have a continuous basement membrane and are connected to each other by a tight junction. There are, however, certain areas within the CNS that lack this barrier: the median eminence, pineal gland, area postrema, subfornical organ, neurohypophysis, supraoptic crest, and choroid plexus. In these particular areas, certain low-molecular-weight substances in the blood will diffuse freely into the interstitial tissue. The common symptoms of nausea and vomiting following injection of contrast media may thus be due to the contrast medium entering the area postrema of the medulla, where it is devoid of a BBB. Diseases that affect the integrity of the BBB may place the patient at increased risk from contrast administration. Brain metastases, for example, have capillaries characteristic of the originating tumor and these do not have such a barrier. Under certain circumstances, contrast media can more readily enter the brain in these regions and may cause seizures. **Seizures after intravenous contrast medium administration are rare in the general population, about 0.01%, but have been reported to be as high as 6–19% in patients with brain metastasis. The risk for seizure development, especially in cases of brain metastasis, may be reduced by an intravenous injection of 5 mg of diazepam prior to the injection of contrast medium.** Should seizures develop, most are also controlled with diazepam.

X. **Urinary iodine concentrations.** For all practical purposes in the normal patient, contrast media are excreted exclusively by the kidney. The peak time for excretion of contrast media is about 3 minutes following intravenous injection. Peak urine iodine concentrations occur approximately 60 minutes after administration. Urine flow rates are increased with the meglumine salts compared with the sodium salts. **The filtration of the nonopaque cation meglumine increases tubule osmolality** and accounts for the finding of lower urinary iodine concentrations with meglumine salts. Sodium ions are partially reabsorbed by the tubule. Thus, some investigators have noted that the sodium salts of diatrizoate or iothalamate give higher urinary concentrations than do comparable doses of the meglumine salt. Nonionic and low-osmolality ionic contrast media produce even higher urinary iodine concentrations.

Within minutes after an intravascular osmotic diuretic is injected, water and sodium excretion increase markedly. The increase in urine flow induced by the osmotic diuresis is associated with an increase in excretion of a wide variety of substances other than water and sodium, and these include potassium, calcium, phosphorus, magnesium, uric acid, urea, and oxalate. Patients with persistent osmotic diuresis are typically volume contracted and hyperosmolar.

- **Endothelial damage and thrombosis.** Contrast media can cause damage to the vascular endothelium, leading to increased permeability of the vessel wall and to the induction of thrombus formation. Clinically, thrombosis has been observed as a complication after intravenous injections of a hypertonic contrast media for phlebography and urography. For this reason, the American College of Radiology has recommended that 60% solutions of ionic contrast media be diluted to 45% when used for phlebography, a practice that may, however, result in the degradation of image quality. Nonionic contrast media, having lower osmolality, reduce endothelial damage, but they also are not anticoagulant. The osmotic gradient across the RBC membrane results in shrinkage, making the cells more rigid with loss of normal deformability. These microcirculatory changes are associated with crenation and aggregation of RBCs, adherence of leukocytes to the vessel wall, and palisading of leukocytes. With ionic contrast media, there is also a sudden decrease in blood coagulation that peaks after about 5 minutes. The depression, however, is transient and returns to normal within 4 hours. A transient decrease in platelet aggregation from contrast medium has been observed. Nonionic contrast media produce a minimal decrease in coagulation, and there has been much discussion about the risks of thrombosis.
- **Hypocalcemia.** Intravascular injections of ionic contrast media produce transient systemic hypocalcemia. Some formulations contain the calcium-binding compounds disodium edetate and sodium citrate that produce significant reductions in ionic calcium. Other factors include the high ionic strength of ionic

contrast media, hemodilution, and direct binding of calcium by the contrast molecule. Hypocalcemia may have potentially serious effects, as this ion is important to cardiac function and rhythm and to the function of all excitable tissue. Hypocalcemia with nonionic contrast media is only related to osmotic dilution and is therefore not very significant.

XIII. **Complement activation.** Complement activation and coagulopathy secondary to contrast media have been described and are believed to be important factors in the development of adverse reactions. Contact activators are known to be present in subendothelial areas that would be subject to release by contrast media-endothelial interactions. Platelet activation also activates the contact system, leading to production of kallikrein and bradykinin. These substances may play a major role in reactions by producing (1) hypotension, (2) smooth muscle spasm, (3) increased capillary permeability, and (4) potentiation of arachidonic acid production favoring the release of vasoactive prostaglandins and leukotrienes. (See Chapter 14 for further discussion.)

XIV. **Pulmonary edema.** In toxicity studies, animals given a lethal dose of contrast medium develop pulmonary edema as well as evidence of CNS damage prior to death. Intravenous injections of contrast media also result in an increase in pulmonary artery pressure (PAP) along with subclinical bronchial spasm, as measured by pulmonary function tests. The effects on PAP have been related to the osmotic effects of the solution for the most part and, less significantly, to the viscosity. **Acute pulmonary edema of noncardiogenic etiology is seen, but rarely in healthy patients** after the intravascular administration of contrast media. In infants, pulmonary edema may be seen after urography, owing to the hypertonicity of the contrast material, particularly because of the higher dose per kilogram of body weight needed. The exact frequency of pulmonary edema occurring in clinical studies is difficult to ascertain, since it is known that interstitial edema may be present without symptoms.

Chapter / 2
Treatment of Acute Reactions to Contrast Media

William H. Bush

I. **General considerations.** Ionic contrast media have been used routinely for intravenous and intra-arterial studies for over 30 years and have proven to be safe and effective. The overall incidence of adverse reactions is 5–8%, with vast majority being of a minor nature, such as nausea, vomiting, a flushing sensation, or a few nonprogressing hives. However, uncommonly (i.e., 1 in 500 to 1 in 1000) a patient will develop a more serious reaction that can prove life threatening, particularly if not treated quickly and appropriately.

More recently, newer contrast agents have been developed. These agents, termed lower osmolar contrast medium or nonionic contrast medium, cause fewer adverse systemic reactions and fewer severe and potentially life-threatening reactions. The reaction rate with these newer nonionic agents is about one-fifth that of conventional standard ionic contrast agents (*AJR* 155:225–233, 1990). Use of these newer nonionic agents has, unfortunately, not completely eliminated serious reactions, and fatal reactions have also occurred (*Radiology* 178:361–362, 1991).

Interestingly, in the largest comparative series (*Radiology* 175:621–628, 1990) when 340,000 patients were reviewed, half of whom had received nonionic contrast and half had received ionic contrast, serious reactions were reduced significantly in the nonionic group, but only a single fatal reaction occurred in each group. The decrease in fatal reactions in the ionic group (compared with earlier studies showing fatal reaction rates of 1 in 75,000 or even 1 in 14,000) likely reflects two important factors in current contrast media usage: patients at higher risk for reaction to contrast are being given the newer nonionic agents; physicians are more cognizant of and better able to treat a serious life-threatening reaction.

Optimally, nonionic contrast could be used for all patients. Unfortunately, the cost of the nonionic agents is very high, 10–20 times more than that of standard ionic agents. Many health care facilities or reimbursement systems are unable to absorb a total conversion to nonionic agents. Although all patients are at some risk from intravascular administration of a contrast medium, patients who are at higher risk to adverse effects can be identified (*Contrast Guidelines*. Chicago: American College of Radiology, 1990). Nonionic contrast agents should be used for these patients as a minimum standard of care (Table 2.1). Patients who have previously had a reaction to contrast should also be premedicated (Table 2.2). Premedication reduces but does not totally eliminate anaphylactoid-type reactions; nonionic contrast plus premedication reduces further the frequency of repeat reactions (*Journal of Allergy and Clinical Immunology* 87:867–872, 1991).

Even if nonionic contrast agents were used universally, however, serious and potentially fatal reactions would occur. Therefore, each physician who is responsible for injection of one of these agents (be it nonionic, lower osmolar, or conventional ionic) must be able to treat effectively the reaction that will occur. Key to treating a reaction effectively are a number of factors, many of which can be defined prior to the injection. **Key factors in treatment of a reaction are:**

- Do not inject contrast in an isolated setting; have help immediately available, and be able to summon additional assistance should a "full-blown" anaphylaxis-like reaction or cardiac arrest occur.

Table 2.1. Current Criteria for the Use of Lower Osmolality Contrast Agents (Nonionic Agents)[a]

1. Previous reaction to contrast (except flushing, heat, nausea, vomiting)
2. History of asthma
3. History of allergy
4. Known cardiac dysfunction (including cardiac decompensation, severe arrhythmias, unstable angina, recent myocardial infarction, pulmonary hypertension)
5. Severe generalized debilitation
6. Certain medical conditions (sickle cell disease; risk of aspiration)
7. Patients manifestly anxious about the contrast procedure
8. Communication problem precluding identification of above risk factors

[a]Adapted from the *Report of the Committee on Drugs and Contrast Media of the Commission on Education of the American College of Radiology,* October 1990.

Table 2.2. Universal Pretreatment Options[a]

Patient Group	Pretreatment
Optimal regimen (3-day)	3-dose, 3-day (32 mg Medrol daily) (last dose 2 hours before examination)
Two-day (12 hours)	2-dose, 2-day (32 mg Medrol, 12 and 2 hours before examination)
Two-day (13 hours)	Prednisone 50 mg orally 13, 7, and 1 hour before examination
	Diphenhydramine 50 mg orally or intramuscularly 1 hour before examination
	Nonionic contrast medium[b]
When 12 hours too long	1-dose (32 mg Medrol, 6 hours before study); nonionic contrast[c]
Immediate emergency	Nonionic contrast (H$_1$-blocker antihistamine, e.g., diphenhydramine, (+) H$_2$-blocker antihistamine, e.g., cimetidine, ranitidine)[b]

[a]*New England Journal of Medicine* 317(14):845–849, 1987.
[b]*Journal of Allergy and Clinical Immunology* 87:867–872, 1991.
[c]*AJR* 150:257–260, 1988.

- Have the equipment and medicines necessary to treat a reaction immediately available.
- Have basic knowledge about the patient.
- Have prior training in treatment of various types of reactions; training in cardiopulmonary resuscitation (CPR) is necessary; training in basic life support (BLS) or advanced cardiac life support (ACLS) protocols is recommended.
- Be able to identify the specific type of reaction so that appropriate, effective treatment can be initiated quickly.
- Early treatment permits use of lower doses of drugs to reverse the reaction, thereby minimizing drug side effects (particularly important with drugs such as epinephrine).

II. **Mechanisms of systemic-type reactions.** Exact etiology of the various contrast reactions is undefined. Several inciting factors may exist, any of which can potentially trigger a reaction, while the final determinant is likely the individual patient's reactivity (see Chapter 14). Stress and fear seem to heighten the chance of a contrast reaction. Histamine is suggested as an important component of the reaction pathway, since the clinical signs and symptoms of an anaphylactoid reaction can be reproduced by intravenous injection of histamine. Complement activation, coagulation system activation, prekallikrein transformation, and generation of bradykinin initiated by contrast contact with the vascular endothelium may be factors. The central nervous system may be a mediator of these adverse reactions via contrast exposure of areas of the brain that normally lack a blood-brain barrier (*Contrast Media: Biological Effects and Clinical Application*. Boca Raton, FL: CRC Press, 1987, pp. 137–150).

III. **Identify the reaction caused by contrast administration.** Reactions include:
- Nausea and/or vomiting.
- Scattered hives; extensive hives without respiratory symptoms.
- Asthma-like symptoms without the cutaneous or vascular manifestations of the generalized anaphylactoid reaction (**bronchospastic reaction**).
- Facial and periorbital edema, nasal stuffiness, sneezing, tightness in throat, shortness of breath, and asthma-like wheezing, usually with developing hypotension and tachycardia (responsive tachycardia may be lacking in a patient taking a beta-adrenergic blocking drug). (This constellation of signs and symptoms identify the so-called **anaphylactoid reaction**.)
- Hypotension (without coexistent asthma-like symptoms), usually with tachycardia unless the patient is taking a beta-adrenergic blocking drug.
- Hypotension and bradycardia (hypotension with associated sinus bradycardia identifies the **vagal reaction**).
- Unconscious, unresponsive, pulseless.
- Major convulsion or seizure.

IV. **Treatment of specific reactions.** Below are listed the recommended treatment plans for contrast reactions defined above. It should be recognized that these treatment outlines are one approach to deal with each specific reaction (*Urology* 35:145–150, 1990). These are not the only methods that can be used to treat contrast reactions, nor are the drugs the only ones that can or should be employed (*AJR* 151:263–270, 1988). **Each physician or department should develop a specific plan or protocol for dealing with the various reactions, update that plan periodically, and be ready to implement it quickly and effectively should a reaction occur** (Table 2.3).

   A. **Nausea and/or vomiting.** If the patient becomes nauseated and begins to vomit, it is usually self-limited and an easily controlled reaction that is not life threatening. However, nausea and vomiting may be one of the earlier signs of an anaphylactoid reaction; therefore, watch the patient closely for other systemic symptoms; be certain that you still have intravenous access.
   1. Stop or slow the rate of contrast injection.

Table 2.3. Systemic Reactions to Contrast Media[a]

Effect	Major Symptoms	Primary Treatment
Vasomotor effect	Warmth Nausea/vomiting	Reassurance
Cutaneous	Scattered hives Severe urticaria	H_1-antihistamines H_2-antihistamines
Bronchospastic	Wheezing	Oxygen Beta-2-agonist inhalers
Anaphylactoid reaction	Angioedema Urticaria Bronchospasm Hypotension	Oxygen IV fluids Adrenergics (IV epinephrine) Inhaled beta-2-adrenergics Antihistamines (H_1- and H_2-blockers) Corticosteroids
Hypotensive	Hypotension	IV fluids
Vagal reaction	Hypotension Bradycardia	IV fluids IV atropine

[a]*Urologic Imaging and Interventional Techniques.* Baltimore: Urban & Schwarzenberg, 1989, pp. 10–18.

   2. Reassure the patient that a serious reaction is not occurring.
   3. If nausea and/or vomiting persists, without other systemic symptoms, administer an antinauseant drug (e.g., Compazine, 5–10 mg IM).
B. **Scattered hives.** If the patient develops a few scattered hives, treatment likely will not be necessary.
   1. Observe the patient closely for other symptoms.
   2. If hives are the only symptom and are troublesome to the patient, administer an antihistamine drug: an H_1-blocker, such as diphenhydramine (Benadryl), 25–50 mg IV, is often effective.
C. **Prominent urticaria.** Urticaria or a diffuse cutaneous reaction may occur independent of a generalized systemic or anaphylactoid reaction. A profound urticarial reaction is effectively treated with an H_2-blocker, such as cimetidine (Tagamet), 300 mg IV, diluted, slowly; or ranitidine (Zantac), 50 mg IV, diluted, slowly (pediatric dose: cimetidine, 5–10 mg/kg IV, diluted, slowly).
D. **Bronchospastic reaction**
   1. Oxygen (nasal prongs or mask, 3 L/min).
   2. Inhalation of a beta-2-agonist drug from a hand-held, single-use metered dose inhaler such as metaproterenol (Alupent), terbutaline (Brethaire), and albuterol (Proventil).

3. Adrenergic drug administered systemically: epinephrine (Adrenalin).
   a. Subcutaneously: 0.1–0.2 mL 1:1000; repeat prn q 10–15 minutes.
   b. Intravenously: 0.1 mL 1:10,000 (slowly IV, 10 μg/min); repeat prn q 5–10 minutes.

E. **Anaphylactoid reaction**
   1. Oxygen (nasal prongs or mask; 3 L/min).
   2. Intravenous fluids (normal saline, rapidly).
   3. Adrenergic drug: epinephrine (Adrenalin) (*JAMA* 251(16):2118–2122, 1984).
      a. If mild to moderate reaction, give subcutaneously 0.1–0.2 mL 1:1000 (pediatric dose: 0.1–0.2 mg, 0.1–0.2 mL, subcutaneously).
      b. If moderate to severe reaction, give IV: small dose, given slowly, 1 mL (0.1 mg) 1:10,000 (pediatric: 0.01 mg/kg IV) (0.1 mL 1:10,000 per minute = 10 μg/min).
      c. Titrate to effect.
      d. Can repeat doses in 5–15 minutes.
   4. Antihistamine.
      a. H_1-blocker: diphenhydramine (Benadryl) 25–50 mg IV.
      b. H_2-blocker: cimetidine (Tagamet) 300 mg IV, diluted, slowly; or ranitidine (Zantac), 50 mg IV, diluted, slowly.
   5. Corticosteroids, such as hydrocortisone (Solu-Cortef), 0.5–1.0 gm IV; or methylprednisolone (Solu-Medrol), 500 mg IV over 30 seconds, or 2000 mg over 30 minutes.
   6. If respiratory symptoms are unresponsive to epinephrine: beta-2-agonist inhaler (metered dose inhaler) (e.g., Alupent, Brethaire, Proventil) 2 or 3 inhalations, do not repeat for 4–6 hours.

F. **Hypotension without bradycardia or bronchospasm.** If the patient is markedly hypotensive, but without bradycardia and without respiratory symptoms of an anaphylactoid reaction, treatment should consist of:
   1. Intravenous fluids: rapidly, hand-pump, 0.9% saline.
   2. Oxygen.
   3. Consider a vasopressor such as dopamine, IV solution, 2–5 μg/kg/min; epinephrine solution, 4–8 μg/min IV.

G. **Vagal reaction.** The combination of prominent sinus bradycardia (<50) and hypotension (systolic pressure <80) identifies the vagal reaction. Recognition of the bradycardia associated with a hypotensive adverse reaction is essential

because treatment is specific and different from that for the anaphylactoid reaction. Treatment involves increasing intravascular fluid volume and reversing the bradycardia.
1. Intravenous fluids rapidly: 0.9% saline—large volume.
2. Atropine: large doses 0.8–1.0 mg IV (pediatric dose: 0.02 mg/kg IV to maximum 0.6 mg). Repeat atropine every 3–5 minutes to a maximum of 3.0 mg (adults) or 2 mg (children). Monitor pulse to determine response to drug (hypotension will not respond immediately, since the atropine is correcting the bradycardia only and is not correcting the vasodilation; intravenous fluids are required to correct the hypotension due to vasodilation) (*Radiology* 121:223–225, 1976).
3. Oxygen.

H. **Unconscious, unresponsive, pulseless.**[a] If a severe reaction has progressed to cardiovascular collapse and ventricular fibrillation or cardiac arrest, standard cardiopulmonary resuscitative measures must be employed. An airway must be established, ventilation initiated, cardiac massage started quickly, and intravenous fluid access established. Cardiac monitoring is then obtained, as defibrillation may be necessary. Although a severe reaction is uncommon, adequate resuscitative skills and equipment must be available and should make death a rare event.
1. Call for help: call a CODE.
2. Check for pulse (femoral, carotid).
3. Sternal "thump": hard.
4. Begin CPR. Use one-way mouth "breather" apparatus to protect yourself, and use external cardiac massage.
5. Oxygen (3 L/min).
6. Intravenous fluids.
7. Defibrillate. Monitor ECG. Shock for ventricular fibrillation, asystole, or ventricular tachycardia in unstable patient: begin with 200 W-sec (J), increase to 300, then to 360.
8. Drugs, such as epinephrine: 1 amp (1 mg) IV, repeat between shocks ×2; atropine: 1 amp (1 mg) IV; and lidocaine: 1 amp (100 mg) IV.

I. **Major convulsion or seizure.** If the patient has a major **convulsion or seizure,**

---

[a]*ACLS: Certification Preparation & A Comprehensive Review.* St. Louis, MO: C. V. Mosby Co., 1987.

Table 2.4. Acute Reactions to Contrast Media: Treatment Outline

Reaction	Treatment
Urticaria	Mild: observation diphenhydramine (Benadryl), 50 mg PO/IM/IV Severe: cimetidine (Tagamet), 300 mg, diluted, slow IV (pediatric: 5–10 mg/kg, diluted, slow IV) ranitidine (Zantac), 50 mg, diluted, IV, slowly

Bronchospasm (isolated)	Oxygen (3 liters/min)
	Beta-2-agonist metered dose inhaler (MDI): 2–3 deep inhalations of metaproterenol (Alupent), terbutaline (Brethaire), or albuterol (Proventil)
	Epinephrine (Adrenalin)
	  Subcutaneously: 1:1000, 0.1–0.2 mL (0.1–0.2 mg)
	    (pediatric: 0.1–0.2 mL subcutaneously)
	  IV: 1:10,000, 1 mL (0.1 mg), slowly
	    (pediatric: 0.01 mg/kg IV)

Anaphylactoid reaction (generalized)	Oxygen (3 L/min)
	IV fluids: 0.9% saline
	Epinephrine (Adrenalin)
	  Subcutaneously: 1:1000, 0.1–0.2 mL (0.1–0.2 mg)
	    (pediatric: 0.1–0.2 mL subcutaneously)
	  IV: 1:10,000, 1 mL, (0.1 mg), slowly (10 µg/min) (pediatric: 0.01 mg/kg IV)
	Antihistamines:
	  H_1-blocker: diphenhydramine, 50 mg IV
	  H_2-blocker: cimetidine (Tagamet), 300 mg, diluted, slowly IV (pediatric: 5–10 mg/kg, diluted, slowly)
	  ranitidine (Zantac), 50 mg, diluted, slowly IV
	Beta-2-agonist metered dose inhaler (MDI) (2 or 3 inhalations): metaproterenol (Alupent); terbutaline (Brethaire); albuterol (Proventil)
	Corticosteroids: hydrocortisone (Solu-Cortef), 0.5–1.0 gm IV
	  methylprednisolone (Solu-Medrol), 500 mg IV over seconds; or 2000 mg over 30 minutes

Hypotension (isolated)	Oxygen (3 L/min) IV fluids (primary therapy): rapidly, 0.9% saline Vasopressor: dopamine: IV solution, 2–5 µg/kg/min epinephrine: IV solution, 4–8 µg/min
Vagal reaction (hypotension and bradycardia)	Oxygen (3 L/min) IV fluids: rapidly, 0.9% saline Atropine: 0.8–1.0 mg IV, repeat q 3–5 minutes to 3 mg total (pediatric: 0.02 mg/kg IV; maximum 0.6 mg to 2 mg total)
Seizure	Diazepam (Valium): 5 mg IV (pediatric: 0.2–0.5 mg/kg IV)

1. This may be the end result of a severe hypotensive episode or vagal reaction.
   a. Protect the patient.
   b. Check history and immediate events.
   c. Check the patient's pulse.
2. If seizuring continues, give an anticonvulsant drug: diazepam (Valium), 5 mg IV (pediatric dose: 0.2–0.5 mg/kg IV) (Table 2.4).

Chapter / 3
Contrast Medium-Induced Nephrotoxicity

Richard W. Katzberg

I. **General considerations**
   A. **Acute renal failure** that is the direct result of contrast administration is a matter of clinical concern. The incidence in published reports varies from zero or a few percent to above 10%, and in some selected series the incidence has been reported to be as high as 50% (*Medicine* 58:270–279, 1979; *Kidney International* 18:540–552, 1980). Newer contrast agents have recently been developed to improve patient safety and imaging quality. This survey examines renal handling of contrast media, mechanisms of contrast enhancement, and questions related to renal toxicity. Methods to minimize contrast medium-induced acute renal failure (CM-ARF) outcomes are presented.

II. **Renal handling of contrast media**
   A. **Dose.** The average adult dose routinely used for conventional urography is 20 gm iodine, which would be in the range of 1–1.5 mL/kg body weight (300 mg I per mL concentration) for most adults, although larger doses of up to 40–80 gm iodine have been advocated. The plasma level of contrast media is dose dependent, and the filtered load (UV) is proportional to the glomerular filtration rate times the plasma level:
   $$GFR = UV/P \text{ or } UV = GFR \times P$$
   where GFR equals glomerular filtration rate in mL/min, U equals amount of contrast material in the urine in mg/mL, V equals volume of urine in mL/min, and P equals plasma concentration of contrast material in mg/mL. The diagnostic quality of an excretory urogram is related to UV, the amount of iodine-containing contrast material excreted by the kidney in a volume of urine that distends the pelvocalyceal system

and ureters (*The Radiology of Renal Parenchymal Diseases.* Philadelphia: W. B. Saunders, 1988, pp. 4 and 5).
B. **Excretion of contrast media.** Contrast material is excreted by glomerular filtration alone and is neither reabsorbed nor secreted by the tubules or collecting ducts.
   1. **The kidney** excretes over 99% of the water-soluble agents administered.
   2. **Extra renal routes of excretion** represent less than 1% of the total amount of water-soluble contrast agent elimination and include liver, bile, small and large bowel, sweat, tears, and saliva.

      Vicarious excretion occurs most commonly in association with renal insult or renal failure. The renal insult can be unilateral, as in obstructive uropathy from calculus disease, and can occasionally be demonstrated in normal patients.
   3. **Peak time** for excretion of contrast media is about 3 minutes following the intravenous injection. Peak urine iodine concentration occurs approximately 60 minutes after administration (*Investigative Radiology* 15(Suppl.): S67–S78, 1980). Urine flow rates increase 4-fold. **Sodium salts of diatrizoate or iothalamate** give higher urinary concentrations than comparable doses of the meglumine salts, since sodium is freely reabsorbable by the renal tubules, whereas meglumine is not reabsorbed by the renal tubules and thus provides an added osmotic force within the tubule lumen to potentiate the diuretic effect of the nonreabsorbable anion (*Clinical Radiology* 21:150, 1970).
C. **Hemodynamic effects.** The administration of hypertonic contrast media either intravenously or directly into the renal arterial circulation induces a biphasic alteration in renal blood flow. There is an initial increase in flow, followed by a decline; the latter is more marked following intra-arterial than intravenous administration. The decrease in renal blood flow has been implicated as a possible mechanism for contrast medium-induced renal toxicity. However, the decrease in renal blood flow is directly related to the osmolality of the contrast agent, and similar changes can be induced by other hypertonic solutions including mannitol (*Investigative Radiology* 18:74–80, 1983).
D. **Tubular effects.** Because molecules of contrast material, like

those of mannitol, are not reabsorbed by the tubular system, they exert an osmotic effect in the tubular lumen, markedly reducing reabsorption of water from the tubules. This increases pressure in Bowman's capsule, which leads to a transient reduction in glomerular filtration rate, filtration fraction, and probably renal perfusion. These effects are markedly attenuated with the use of low-osmolality contrast agents.

Within minutes after an intravascular osmotic is injected, water and sodium excretion increase markedly. Much of the diuretic action occurs because of the inhibition of sodium and water reabsorption in the proximal tubules along with inhibition of sodium and water transport in the loops of Henle. During brisk osmotic diuresis, the distal tubule and collecting duct fail to recapture the increased sodium and water load delivered into the early distal tubule. Increases in intraluminal flow to the distal portion of the loop of Henle lead to a decrease in whole-kidney glomerular filtration rate, which is a function of the glomerulotubular feedback mechanism. This is, in part, related to the marked increase in proximal tubular pressure occurring with increased flow rates within the nephron (*American Journal of Physiology* 189:323–328, 1975). The increase in urine flow produced by the osmotic diuresis is associated with an increase in excretion of a wide variety of substances other than water and sodium; these include potassium, calcium, phosphorus, magnesium, uric acid, urea, and oxalate. Patients with persistent osmotic diuresis are typically volume contracted and hyperosmolar (*New England Journal of Medicine* 291:714–720, 1974). Thus, **the contrast media examination can lead to significant patient dehydration,** which is to be avoided.

E. **Renal enhancement.** The total quantity of contrast medium excreted is more important for opacification than concentration, though both factors are significant. Renal enhancement can be improved by (1) increasing the dose of the contrast material injected, (2) dehydrating the patient prior to the contrasted examination, and (3) utilizing low osmolar agents. As the standard dosage of contrast materials has generally increased in the past decade, the need for the radiodensity-enhancing value of dehydration has decreased. Dehydration is a risk factor for contrast material-induced nephrotoxicity and should be corrected either prior to or immediately following the examination.

II. **Nephrotoxicity. Contrast medium-induced acute renal failure (CM-ARF)** is defined clinically as a rise in serum creatinine level of at least 1.0 mg/dL, occurring within 48 hours after administration of the contrast media. Oliguria (urine volume of less than 400 mL/day) is observed in approximately 75% of patients. Nonoliguric renal failure occurs in the other 25%.

Predisposing factors for CM-ARF include preexisting renal insufficiency (serum creatinine level of 1.5 mg/dL or greater), diabetes mellitus, dehydration, cardiovascular disease and use of diuretics, advanced age ($\geq$ 70 years), multiple myeloma, hypertension, and hyperuricemia (Table 3.1) (*Medicine* 58:270–279, 1979; *Kidney International* 18:540–552, 1980).

A significant but transient proteinuria develops after renal angiography in animals and in humans with the use of hypertonic agents and with the nonionic agent metrizamide. Abnormal enzymuria has also been reported after both the intra-arterial and intravenous injections of contrast materials (*Contrast Media.* Stuttgart: Georg Thieme Verlag, 1983, pp. 30–36).

Hypertonic contrast agents produce vacuolization in the cytoplasm of the renal proximal tubular cells. This has been termed "**osmotic nephrosis**" and is seen most commonly after the administration of contrast materials in patients with preexisting renal insufficiency. However, similar changes have been found after urography with metrizamide, and in a small number of patients studied, osmotic nephrosis was seen in some patients after urography with Hexabrix and iopamidol. Since these are low-osmolality agents, chemotoxicity rather than osmolality may be a factor. **These changes have not been shown to have a direct relationship to significant renal toxicity and are clearly nonspecific.**

IV. **Hypothetical mechanisms of contrast medium-induced acute renal failure.** The mechanisms leading to the development of CM-ARF following injections of intravenous contrast medium

Table 3.1. Risk Factors for CM-ARF

Preexisting renal insufficiency
Diabetes mellitus
Dehydration
Cardiovascular disease and diuretics
Advanced age ($\geq$70 years)
Multiple myeloma
Hypertension
Hyperuricemia

are not completely understood. Some of the hypotheses that have been suggested are hemodynamic alterations (direct effects of contrast media on renal hemodynamics), intratubular obstruction, direct tubular cell injury, the prerenal effects from hypotension and/or dehydration, and immunologic mechanisms (Table 3.2) (*Kidney International* 18:540–552, 1980; *American Journal of Physiology* 252:F246–F255, 1987).

V. **Radiologic findings in contrast medium-induced acute renal failure.** The nephrogram in CM-ARF becomes dense immediately following the contrast material injection and remains so for a prolonged time, sometimes persisting for >24 hours (Fig. 3.1). This is observed in approximately 75% of patients. A second time-density pattern that is seen in approximately 25% of patients is a nephrogram that becomes increasingly dense in the time interval during the contrast material-enhanced study. The kidneys are bilaterally enlarged and smooth in outline. Opacification of the pelvocalyceal system is poor, and the collecting system may be effaced by the surrounding interstitial edema. The echogenicity by ultrasonic evaluation is normal to diminished in the medullary region and normal to increased in the cortical region.

VI. **Treatment of contrast medium-induced acute renal failure** may include conservative measures or dialysis, depending on the severity of renal impairment and resulting complications. The conservative approach is facilitated by weighing the patient daily, accurately recording fluid intake and output, and frequently (at least 3 times/week) measuring serum electrolytes, BUN, Cr, calcium, and phosphate. Fluid management requires accurate clinical judgment of intravascular volume. Volume depletion may contribute to acute renal failure by decreasing renal perfusion and must be corrected. Protein intake is limited to approximately 0.5 gm/kg/day to decrease nitrogenous waste production while enough calories (35–50 kcal/kg/day) are provided to avoid catabolism. Blood pressure is frequently evaluated and corrected with volume expansion, depending on the patient's intravascular volume status. Elevated serum phos-

Table 3.2. Hypothetical Mechanisms of CM-ARF

Hemodynamic alterations
Intratubular obstruction
Tubular cell damage
"Prerenal" hypotension; dehydration
Immunologic factors

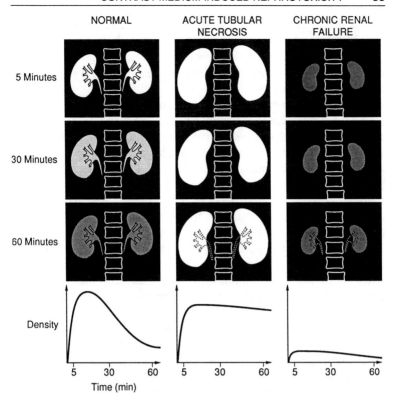

Figure 3.1. The nephrographic patterns and time-density curves (*lower row*) with excretory urography in normal, acute tubular necrosis, and chronic renal failure kidneys. The most commonly encountered pattern with CM-ARF is depicted by the "acute tubular necrosis" sequence (*middle diagram*).

phate levels are treated by decreasing intestinal absorption of phosphate with aluminum hydroxide-containing antacids. Hyperkalemia is treated with dietary restriction and potassium-binding resins. Metabolic acidosis is commonly observed but, if mild, does not require therapy. More marked acidosis is corrected with sodium bicarbonate, 300–600 mg PO tid.

Dialysis for CM-ARF is indicated when **severe hyperkalemia, acidosis, or volume overload** cannot be controlled by conservative measures. Signs and symptoms of uremia and the need for aggressive nutritional therapy may also necessitate institution of dialysis.

Management of the recovery phase of CM-ARF is accomplished by careful monitoring of serum electrolytes, volume status, and urinary fluid and electrolyte loss. As with obstructive nephropathy, a diuretic phase may occur during recovery and is often physiologic. With time, tubule function returns and the kidney once again regains its ability to concentrate the glomerular filtrate. An overwhelming majority of contrast medium-induced acute renal failures are mild and self-limited with a good prognosis.

VII. **New contrast agents.** Research has aimed at the development of new contrast media that reduce toxicity, unpleasant side effects, and adverse reactions while improving diagnostic quality. The new contrast media are of three basic types: nonionic monomers, monoacidic dimers, and nonionic dimers. These agents have one major characteristic in common: a markedly lower osmolality. Clinical usage has shown fewer side effects, less toxicity, and equivalent or better diagnostic opacification than with the use of the current-ratio agents.

Acute toxicity (LD_{50}) in mice for new agents shows an LD_{50} of 13–25 gm iodine per kg compared with 7–10 gm iodine per kg with the current ionic monomers. Low-osmolality media produce significantly higher urinary iodine concentrations, less urine volume (less diuretic effect), and similar total iodine excretion rates, compared with the high osmolar contrast agents. Extensive clinical investigations in Japan demonstrate a significant decrease in general side effects (*Radiology* 175:621–628, 1990). The relative advantages of low-osmolality media in lowering the likelihood of nephrotoxicity is still controversial, however (*New England Journal of Medicine* 320:149–153, 1989).

VIII. **Prevention of contrast medium-induced acute renal failure.** In patients with normal renal function the risk of CM-ARF is very low and should not lead to a choice of inferior diagnostic methods. If the radiologist is aware of the risk factors involved in the development of CM-ARF but the contrast study is deemed necessary, precautions can be undertaken to reduce the incidence. Choices available are shown in Table 3.3 and include **careful patient selection** (avoid those with risk factors), **no dehydration** in preparation for the contrasted study, **increasing the interval between contrast medium examinations,** and

Table 3.3. Prevention of CM-ARF

Select patient carefully: avoid risk factors.
Do not dehydrate!
Increase interval between contrast-enhanced examination.
Select type and dose of contrast medium carefully.

selecting contrast media of lower osmolality with lower dose when significant risk factors have been identified in the patient.

Fortunately, most episodes of CM-ARF are self-limited without serious sequelae. Since so many of the physiologic actions of contrast media are related to osmolality, it is my opinion that the introduction of low-ionic and nonionic water-soluble contrast agents will lead to lower risk.

Chapter / 4

Imaging in the Patient with Azotemia

Arthur J. Segal

The workup of the patient with azotemia is designed to sort through the diagnostic possibilities in a quick and cost-effective manner, then to identify and confirm the cause of azotemia. There is usually an incentive to find a diagnostic pathway that does not involve the use of systemic contrast media. However, under certain circumstances, its use may be necessary. In these instances, the type and amount of contrast will be important.

I. **Diagnostic tools**
  A. **Plain film of the abdomen** is usually the first imaging examination. It is quick, cost-effective, and relatively ubiquitous and can provide information regarding the location, size, and contours of the kidneys as well as the presence and distribution of calculi or gas collections within the urinary tract. Underlying metabolic or metastatic bone disease may be revealed by examination of the regional bones, and occasionally, there may be evidence of metastatic disease at the lung bases.
  B. **Ultrasonography (US).** To detect dilatation of the upper collecting system in the presence of obstruction, particularly in the patient with azotemia, is usually possible. With US, one may be able to assess renal size and contours, determine the location and/or cause of the obstruction, and detect the presence of mass(es) or calculi. Evaluation of the renal veins and inferior vena cava for thrombus (with or without tumor) is possible. The use of Doppler technique is currently under investigation as a means of assessing renal artery inflow.

  Ellenbogen et al. (*AJR* 130:731–733, 1978) showed that "when obstruction is the sole clinical question," US could detect hydronephrosis with a sensitivity of 98% (false negatives—2%). However, given the same question, it was only

74% specific (false positives—26%). Other authors have established a false positive rate that is somewhat lower, 8–26% (B. L. McClennan. *Radiologic Clinics of North America* 17:197–211, 1979; L. B. Talner et al. *Urologic Radiology* 3:1–6, 1981).

False positive diagnoses (Table 4.1) may be seen in association with and include a large extrarenal pelvis, peripelvic cyst, multicystic dysplastic kidney, polycystic kidneys, renal artery aneurysm, distended bladder, postoperative renal pelvic anatomy, pregnancy, reflux, and other possibilities. False negative results (Table 4.2) may occur in the presence of errors that result from the examiner (interpretive), equipment (not state of the art), physical factors (e.g., obesity, dehydration, ileus), and pathologic factors (e.g., calculus disease and tumor) (Curatola et al. *Journal of Urology* 130:8–10, 1983; E. S. Amis et al. *Urology* 19:101–105, 1982). Also in this latter group are the problematic 4–5% of patients who present with nondilated or minimally dilated obstructive uropathy (Spital et al. *Urology* 31:478–482, 1988; Maillet et al. *Radiology* 160:659–662, 1986) usually in the setting of malignancy or retroperitoneal fibrosis.

Curatola et al. (*Journal of Urology* 130:8–10, 1983) showed

Table 4.1. Causes of False Positive Diagnoses of Hydronephrosis by US

Large extrarenal pelvis
Peripelvic cyst
Multicystic kidney
Polycystic kidneys
Renal artery aneurysm
Distended urinary bladder
Postoperative renal pelvic anatomy
Pregnancy
Reflux
Other[a]

[a]See E. S. Amis et al. *Urology* 19:101–105, 1982, for a more complete list.

Table 4.2. Causes of False Negative Diagnoses of Collecting System Obstruction by US[a]

Interpretive—expertise of examiner
Equipment—state of the art or outdated?
Physical factors—e.g., obesity, dehydration, ileus
Pathologic factors—e.g., calculus disease, tumor
Nondilated obstructive uropathy

[a]See E. S. Amis et al. *Urology* 19:101–105, 1982, for a more complete list.

that in patients with renal failure (serum creatinine >2.0 mg), an abnormal US increased the probability of surgery from 0.22 to 0.81, and a negative study decreased the probability from 0.22 to 0.013. Ritchie et al. (*Radiology* 167:245–247, 1988) took a group of patients with azotemia and separated them into high- and low-risk populations based on the clinical likelihood of having postrenal urinary obstruction. US documented a prevalence of only 1% hydronephrosis in the group with a low pretest probability of renal obstruction.

On US, it may be found that the collecting systems are not dilated and that the kidneys are more echogenic than the liver. When this is seen in association with bilaterally small kidneys (8 cm or less; normal 9–13 cm) and azotemia, end-stage renal failure is presumed, and the specific etiology, if not clinically obvious, may only be obtainable with histologic evaluation.

There are a large number of disease states associated with normal or large kidneys and azotemia. In these cases, the renal failure is relatively acute and may be reversible. Intrarenal causes of **acute renal failure** (Table 4.3) are extensive but include acute tubular necrosis, glomerulonephrosis, bilateral cortical necrosis, papillary necrosis, interstitial renal disease (e.g., acute pyelonephritis, etc.), and vascular diseases (e.g., artery or vein occlusion). It may not always be apparent at the time of acute presentation whether or not one is seeing a patient with a truly acute renal process or a more chronic disease that is clinically presenting acutely. The intrarenal causes of **chronic renal failure** (Table 4.4) include adult polycystic disease, glomerulonephritis, chronic pyelonephritis, renal tubular acidosis, papillary necrosis, diabetic nephropathy, malignant hypertension, amyloidosis, and Wegener's granulomatosis (J. A. Becker. In:

Table 4.3. Intrarenal Causes of Acute Renal Failure[a]

Acute tubular necrosis
Glomerulonephrosis
Bilateral cortical necrosis
Papillary necrosis
Interstitial renal disease
Vascular diseases

[a]See J. A. Becker. In: H. Pollack, ed. *Clinical Urography*. Philadelphia: W. B. Saunders, 1990, pp. 2596–2597, for a more complete list.

H. Pollack, ed. *Clinical Urology.* Philadelphia: W. B. Saunders, 1990, pp. 2596–2597).

There are acute and chronic causes of postrenal failure with collecting system obstruction. **Acute renal failure** may be due to calculus-induced ureteric obstruction, bladder outlet obstruction (e.g., prostatism), iatrogenic causes following surgery or cystoureteroscopy, hematuria with clot colic, acute ureteropelvic junction obstruction, and other conditions (Table 4.5). **Chronic renal failure** may be due to bladder outlet obstruction (e.g., prostatism), malignant obstruction of the ureters (e.g., retropertoneal, renal, ureteral, bladder, prostate, cervix) retroperitoneal fibrosis, ureteral strictures (e.g., following radiation or surgery), papillary necrosis with sloughed papillae causing ureteric obstruction, postinfectious obstruction (e.g., tuberculosis, xanthogranulomatous pyelonephritis), functional obstruction (e.g., primary megaloureters), endometriosis, and a variety of other conditions (Table 4.6). **Note:** Several of these conditions are less likely to occur in patients with two functioning kidneys (see footnotes in Tables 4.5 and 4.6).

It is important to emphasize that pretest probability is

Table 4.4. Intrarenal Causes of Chronic Renal Failure[a]

Adult polycystic disease
Glomerulonephritis
Chronic pyelonephritis
Renal tubular acidosis
Papillary necrosis
Diabetic nephropathy
Malignant hypertension
Amyloidosis
Wegener's granulomatosis

[a]See J. A. Becker. In: H. Pollack, ed. *Clinical Urography.* Philadelphia: W. B. Saunders, 1990, pp. 2596–2597, for a more complete list.

Table 4.5. Causes of Acute Azotemia with Collecting System Obstruction

Calculi
Bladder outlet obstruction (e.g., prostatism)
Iatrogenic (postsurgical or cystoureteroscopy)
Hematuria with clot colic[a]
Acute ureteropelvic junction obstruction[a]
Other

[a]Not a common cause of azotemia unless there is bilaterally severe disease or one kidney is absent (anatomically or functionally).

Table 4.6. Causes of Chronic Azotemia with Collecting System Obstruction

Bladder outlet obstruction (e.g., prostatism)
Malignant obstruction of ureters or bladder
Retroperitoneal fibrosis
Ureteral strictures (e.g., radiation, postsurgical)
Papillary necrosis with sloughing
Postinfectious obstruction[a] (e.g., tuberculosis, xanthogranulomatous pyelonephritis)
Ureteropelvic junction obstruction[a] (congenital or acquired)
Functional obstruction[a] (e.g., primary megaloureters)
Endometriosis[a]
Other

---

[a]Not a common cause of azotemia unless there is bilaterally severe disease or one kidney is absent (anatomically or functionally).

crucial to the value of utilizing the classical ultrasonographic criteria for renal echogenicity. If patients are examined without knowledge of clinical data, it is possible to use renal parenchymal echogenicity as an indicator of renal disease. In adults, when renal echogenicity is **greater** than that of the liver, detection of renal disease is specific (96%) and has a positive predictive value (67%) in spite of its relative insensitivity (20%) (Platt et al. *AJR* 151:317–319, 1988). In spite of this, even when pretest probability of renal disease is high, it should be noted that most cases of medical renal failure will have kidneys of normal echogenicity J. A. Becker. In: H. Pollack, ed. *Clinical Urography*. Philadelphia: W. B. Saunders, 1990, p. 2607).

C. **Computed tomography (CT).** US is usually preferred to CT in the setting of azotemia. However, sometimes, technical problems preclude accurate imaging with US, and in that setting, CT (Table 4.7) may provide valuable information about renal size, location, contours, calculi, or masses. Even without intravenous injection of contrast media, it is frequently possible to determine renal parenchymal thickness as well as the degree of pelvicalyceal and ureteral dilatation (B. L. McClennan. *Radiologic Clinics of North America* 17:197–211, 1979). If IV contrast can be injected, the aorta and renal arteries and veins can be evaluated with dynamic CT. False positive diagnoses can occur in the presence of peripelvic cysts, subcapsular hematomas (S. E. Amis et al. *Journal of Computer Assisted Tomography* 6:511–513, 1982), and dilatation without obstruction. False negative examinations may occur in the setting of nondilated obstructive uropathy (Spital et al. *Urology* 31:478–482, 1988). However,

Table 4.7. New Patient with Evidence of Azotemia

Initial evaluation
Supine radiograph
US
Secondary evaluation
CT (when US is inadequate or incomplete)

some causes of obstruction are more easily detected by CT than by other modalities. This is particularly true for calculi, particularly if they are composed of uric acid or are hidden by the bony pelvis, sacrum, or overlying bowel gas (A. J. Segal et al. *Radiology* 129:447–450, 1978). CT can readily detect obstruction caused by tumor, whether intrarenal, intraluminal, or retroperitoneal. In some cases, this may be hard to distinguish from retroperitoneal fibrosis on this examination alone. Occasionally, inflammatory disease may be found to cause obstruction in the patient with azotemia.

Patients occasionally have a CT scan 1–4 days following another diagnostic examination during which they received an intravascular contrast load. In some cases, there may be retained contrast with increased density of the renal parenchyma. Love et al. (*Radiology* 172:125–129, 1989) have shown that these patients have developed interval renal failure or worsening of preexisting renal failure (possibly "contrast nephropathy").

D. **Nuclear imaging.** In the patient with azotemia, nuclear imaging is usually not a first-line diagnostic test. However, there may be exceptions to this rule when evaluating a patient following renal transplantation and in trying to distinguish between obstruction and rejection. Furthermore, dynamic and static imaging techniques may be important in assessing renal size, determining relative renal function (i.e., right versus left), and detecting the presence of obstruction. Diagnosis of a urine leak or urinoma is also possible.

A variety of compounds have been used to attain this information. Orthoiodohippurate (1–131 or 1–123) has been effective in renography because Hippuran is excreted by tubular secretion and, in the presence of normal renal function, is rapidly cleared from the blood. Renography is a graphic representation of the dynamic excretion pattern of the hippurate compound from each kidney and can be of help in detecting obstruction, whether unilateral or bilateral. Technetium (Tc)-99m-DTPA (diethylenetriamine pentaacetic acid) is excreted by glomerular filtration and, compared with hippurate, results in improved imaging of the kidney and ureter. Unfortunately, detail worsens significantly in the presence of renal failure. Both Tc-99m-GHA (glucoheptonate) and Tc-99m-DMSA (dimercaptosuccinic acid) have even more limited roles in the imaging of the patient with azotemia but have been used to estimate residual functioning renal parenchyma following resolution of obstruction (S. C. Scharf and M. D. Blaufox. Radionuclides in the Evaluation of Urinary Obstruction. In: *Nuclear Medicine in Clinical Urology and Nephrology.* New York: Appleton-Century-Crofts, 1985, 191–206).

More recently, Tc-99m-MAG$_3$ (mercaptoacetyltriglycine) has been used in assessing patients with impaired renal function following renal transplantation. It is a renal tubular agent that is cleared with greater efficiency and with improved target organ-to-background distinction. (J. R. Taylor et al. *Clinical Nuclear Medicine* 15:371–378, 1990) when compared with those agents that are excreted by glomerular filtration.

E. **Excretory urography (EU).** In renal failure, EU is not frequently desired because (1) contrast can result in further deterioration of renal function and (2) there is usually insufficient concentration of contrast media to provide collecting system or renal parenchymal detail. However, with milder degrees of renal compromise (less than 1.5 mg/dL in the diabetic patient and less than 4 mg/dL in many other nonobstructive causes of renal disease), the EU may reveal renal size and contours as well as the location of calculi and mass(es). In the presence of mechanical obstruction, delayed films are frequently required to reveal urinary tract detail and to define the level and possible cause of obstruction.

F. **Pyelography (retrograde/antegrade).** When obstruction is diagnosed or suspected but the cause and location have not been determined, direct opacification of the upper collecting system can be accomplished with retrograde or antegrade pyelography. Retrograde examination is usually preferred because it is slightly less invasive; however, direct injection of the upper collecting system may be required if the retrograde examination is not technically possible or is inadequate. This examination can be accomplished by utilizing US, fluoroscopy, or CT. If collecting system obstruction is present, a double-J catheter can be inserted following either

retrograde or antegrade pyelography; alternatively, a percutaneous nephrostomy can also relieve upper tract obstruction.
G. **Magnetic resonance imaging (MRI).** MRI may be utilized to define renal anatomy, pathology and, with intravenous contrast media, function. Even without exogenous contrast media, T1- and T2-weighted sequences permit the evaluation of flow and the detection of tumor or thrombus in major vessels. Its high cost, lack of universal availability, difficulty in examining many sick patients (because of motion, life support equipment with ferrous components, or claustrophobia), and lack (in most cases) of superior information to CT and US have made this a relatively minor modality in the workup of the patient with azotemia.
H. **Arteriography.** Arteriography may be necessary when the cause of renal failure is thought to be due to arterial obstruction. This could result from any diffuse vascular disease (e.g., arteriosclerosis, vasculitis), emboli, or any other cause of bilateral renal arterial hypoperfusion in a patient with two functioning kidneys. (R. L. Clark. In: C. E. Putman and C. E. Ravin, eds. *Textbook of Diagnostic Imaging*. Philadelphia: W. B. Saunders, 1988, pp. 1233–1244). Patients with one functioning kidney, such as may be seen with congenital or surgical absence, can also present with azotemia, as can those patients who have two kidneys, one of which has preexisting disease, when there is associated renal vascular obstruction or arterial hypoperfusion.
I. **Venography** may be necessary when the cause of renal failure is thought to be renal vein thrombosis (RVT). In the adult, causes include membranous glomerulonephritis, hypercoagulability states, extension of ovarian vein thrombosis, and extension of or compression by carcinoma or lymphoma. In the infant, dehydration, hypotension, sepsis, and sickle cell disease. (H. Z. Mellins. In: H. Pollack, ed. *Clinical Urography*. Philadelphia: W. B. Saunders, 1990, pp. 2119–2126). As with arterial hypoperfusion, azotemia due to RVT is likely if it is bilateral. Unilateral RVT should only produce renal failure if it occurs in a patient with a single kidney or, alternatively, in a patient with two kidneys, one or both of which have preexisting poor function.
J. **Biopsy.** If, after diagnostic workup, the exact nature of the renal failure is not determined but is thought to be due to renal parenchymal disease, percutaneous renal biopsy can be

performed, usually under US guidance; however, CT provides an excellent alternative, especially in cases that are not clearly imaged by US (J. J. Cronan. In: H. Pollack, ed. *Clinical Urography*. Philadelphia: W. B. Saunders, 1990, pp. 2615–2616).

   K. **Cystography** is not usually considered in the workup of the patient with azotemia. There are, however, two circumstances in which it may be used. The first occurs in the setting of the oliguric or anuric patient with azotemia where there is known, or there is suspicion of, previous trauma (either extrinsic or iatrogenic). A bladder tear may result in a persistent leak of urine, with reabsorption of urine producing the picture of decreased urine output in the setting of progressive azotemia. Secondly, it may be necessary to document suspected vesicoureteral reflux in patients prior to renal transplantation.

II. **Generic causes of azotemia**
   A. **Collecting system obstruction**
   B. **Arterial obstruction or insufficiency**
   C. **Venous obstruction**
   D. **Reabsorption of extravasated urine**
   E. **Renal parenchymal disease**

III. **Primary diagnostic pathways.** The most typical and primary diagnostic pathways (represented by *thick arrows* in Fig. 4.1) arise when a patient presents with laboratory and/or clinical evidence of renal failure. As indicated above, a simple and informative examination is the **supine radiograph.** Immediately following this, **ultrasonography** will be performed. At the completion of these two examinations, which are rapid, relatively inexpensive, and noninvasive, a moderate amount of diagnostic data have been accumulated. If one or both of these examinations are inadequate and/or incomplete, **CT** without and, depending on the degree of azotemia, with IV contrast enhancement will be of help in evaluating the abdomen and kidneys (Table 4.7).

   A. **No collecting system obstruction** (Table 4.8). If the kidneys are small, percutaneous **renal biopsy** is of most help in achieving histologic confirmation of the underlying pathologic process. If they are normal or large, the diagnostic possibilities include metabolic, infiltrative, or other renal parenchymal disease as well as obstruction of the arterial or venous system. If metabolic or parenchymal disease is suspected, renal biopsy would be most effective. If the answer

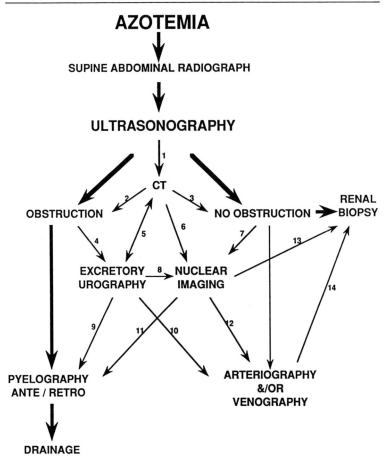

Figure 4.1. Secondary diagnostic pathways (see text for description). *Thick arrow* represents a primary pathway. *Thin arrow* represents a secondary pathway.

is "probably infiltrative neoplastic," renal biopsy may be necessary, but additional diagnostic information may be required with the use of **CT** (or MRI). If there is suspicion of obstructed arterial flow to the kidneys, this can be screened with **nuclear imaging (renography)** or, possibly, **color Doppler US** and more thoroughly evaluated with **arteriography**. With arterial digital subtraction angiography (**DSA**) equipment, the examination can be performed by utilizing a minimum of contrast in the patient with azotemia. Similarly, if the question of renal vein thrombosis or obstruction is considered, **renal venography** should be performed, again, if

Table 4.8. Azotemia without Collecting System Obstruction[a]

Small kidneys
  Biopsy—frequently required
  EU—only occasionally of help (e.g., chronic pyelonephritis in the setting of mild azotemia)
Normal or enlarged kidneys
  Parenchymal disease—biopsy frequently required
  Neoplastic disease—biopsy usually required
  Arterial obstruction
    US (color Doppler)
    Renogram
    Dynamic CT
    MRI—uncommonly used to make this diagnosis but possible
    Arteriography (*most definitive*)—DSA if possible
  Venous obstruction
    Dynamic CT
    MRI
    Venography—DSA if possible

[a]Patient has already had conventional US and/or CT *without definitive diagnosis*.

possible, by utilizing DSA technology. **Note:** Not all patients are candidates for DSA. They must be cooperative and hold still (usually with breath-holding during the sequential exposures following an injection) to allow for precise subtraction.

B. **Collecting system obstruction** (Table 4.9). Typically, direct opacification of the upper urinary tract will be necessary at this point. In many institutions, it is a urologic decision as to how this is done. If **retrograde pyelography** and/or double-J **stenting** is desired, it is most often performed by a urologist. If **antegrade pyelography** or **percutaneous nephrostomy** is preferred, this is usually performed by a radiologist. Temporary or more long-term **drainage** is possible by utilizing stenting or nephrostomy.

Although the level of obstruction can be determined with the above-described techniques, the diagnosis is not always clear at this point. If the **obstruction** is **intraluminal,** the most common possibilities in the patient with azotemia would include one or more calculi versus intraluminal transitional cell carcinoma (TCC). In that setting, if the diagnosis could not be made with **US** and **supine radiography,** then calculi should be diagnosable by **CT,** and tumor can be diagnosed with either **CT** and/or **ureteroscopy** or **ureteral brush biopsy.**

If the **obstruction** is **extraluminal,** then, excluding iatrogenic possibilities in the postoperative patient, the most likely etiology is neoplastic obstruction of the ureter(s). Occasionally, other etiologies may be considered, such as retroperitoneal fibrosis, inflammatory disease, and endometriosis. This can frequently be diagnosed or suspected by **CT** or **MRI**. If a preoperative diagnosis is desired, percutaneous **needle biopsy** utilizing **US** or **CT** can be performed.

7. **Secondary diagnostic pathways.** Secondary diagnostic pathways are represented by the *thinner black arrows* in Figure 4.1. They represent those additional diagnostic tests that may be required because one of the major pathways was unavailable or proved unsuccessful. Alternatively, the diagnosis may be discovered while another problem or the wrong diagnostic consideration is investigated.

If one is interested in a simpler chart, see Fig. 4.2. This offers fewer pathways and is less complex to view. However, in either case, one may use the description of pathways to try to understand the variety of circumstances that may lead to the use of these many modalities. To the student, my advice is to try to understand the pathways but discuss your proposed choice with a radiologist who is familiar with the variety of possibilities and who has the experience to sort through this maze and choose the best option for an individual patient. **"WHEN IN DOUBT, ASK FOR ADVICE!"**

A. **Pathway 1 (US → CT).** CT is used when US is inconclusive.
B. **Pathway 2 (CT → obstruction).** When obstruction is found by CT, luminal detail including the level and the cause of obstruction is required; then resume the major pathway toward pyelography.

Table 4.9. Azotemia with Collecting System Obstruction (Defined by US and/or CT)

Luminal obstruction
   Retrograde pyelography—double-J stent may be placed
   Antegrade pyelography—usually with percutaneous nephrostomy
   CT—confirm calculus or tumor
   US—confirm calculus or tumor
   Ureteroscopy/ureteral brush biopsy—with histologic confirmation of tumor
Extraluminal obstruction
   CT—define level and/or possible cause of obstruction
   US—define level and/or possible cause of obstruction
   Biopsy—histologic diagnosis guided by CT or US

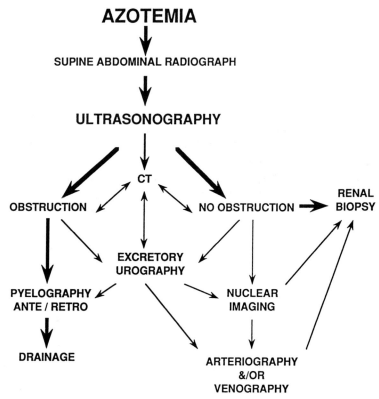

Figure 4.2. Flow chart similar to Figure 4.1 but with fewer pathways. *Thick arrow* represents a primary pathway. *Thin arrow* represents a secondary pathway.

- **C. Pathway 3 (CT → no obstruction).** When no obstruction is defined by CT and the kidneys are small, resume the major pathway to biopsy. This pathway should be similarly pursued when parenchymal, nonneoplastic disease is suspected.
- **D. Pathway 4 (obstruction → EU).** When obstruction is defined by US and the cause is presumed to be intraluminal, EU may be used if azotemia is not severe enough to present a contraindication to systemic contrast media.
- **E. Pathway 5 (CT ↔ EU).** In the setting where US is not sufficiently definitive, use either (1) or (2).
  1. **CT → EU.** CT may define obstruction; if azotemia is not too severe and additional luminal detail is required, EU can be performed.
  2. **EU → CT.** If EU defines the level of obstruction, but there is a question of either intraluminal calculus, clot, or

cancer, or additional extraluminal detail is required to define the cause of obstruction, CT can be performed.

F. **Pathway 6 (CT → nuclear imaging).** If there is inadequate information to define the cause of azotemia on US or CT and a screening examination is required to rule out luminal or vascular obstruction, renography can be performed.

G. **Pathway 7 (no obstruction → nuclear imaging).** The renogram can be used as a screening examination when US has likely excluded luminal obstruction and there is suspicion of renal vascular impairment.

H. **Pathway 8 (EU → nuclear imaging).** When US and EU have likely excluded luminal obstruction, renography can be used to screen for vascular obstruction.

I. **Pathway 9 (EU → pyelography).** If EU is not definitive and improved luminal detail is required for diagnosis, pyelography can be performed. The possibility of drainage following this diagnostic examination can then be considered.

J. **Pathway 10 (EU → angiography).** When an absent or abnormal nephrogram on EU leads to the strong suspicion of vascular obstruction as the cause of renal failure, angiography should be considered.

K. **Pathway 11 (nuclear imaging → pyelography).** Following the detection of collecting system obstruction by nuclear imaging, pyelography may be required to define better the level and cause of obstruction. The possibility of drainage following this diagnostic examination can then be considered.

L. **Pathway 12 (nuclear imaging → angiography).** If the nuclear imaging study supports the diagnosis of vascular obstruction, angiography may be required for confirmation and improved vascular detail.

M. **Pathway 13 (nuclear imaging → renal biopsy).** There is no support for either luminal or vascular obstruction on nuclear imaging, and histologic confirmation of disease can be achieved by renal biopsy.

N. **Pathway 14 (angiography → renal biopsy).** If vascular obstruction or specific vascular disease is not confirmed by angiography, histologic confirmation of disease can be achieved by renal biopsy.

Chapter / 5
Urography, Cystography, and Urethrography

Bernard F. King
Robert R. Hattery

The number of intravenous urograms (IVUs) has decreased significantly in the past 15 years, in part, because of alternative imaging modalities (ultrasound and computed tomography (CT)). Magnetic resonance (MR) has limited application for imaging of the upper urinary tract, but specific applications for evaluation of the lower genitourinary tract are evolving. Intravenous urography continues to be the primary modality of choice for visualization of pathology affecting the pelvocalyceal system and the ureters, for assessment of renal anatomy, and for evaluation of stone disease and infection. Although ultrasound, CT, and MR are routinely employed for evaluation of the pelvis, cystography and urethrography are important imaging techniques for the evaluation of the bladder and urethra. It is vital to employ optimum urographic techniques to achieve the highest quality examinations tailored to solve diagnostic problems.

I. **Intravenous urography**
   A. **Goals**
      1. **Plain film evaluation** of the urinary tract is the initial part of the IVU. Detection of occult abnormalities of the spine and pelvis, visualization of soft tissue pathology, and assessment of stone disease are dependent on high-quality plain film technique. Following the administration of contrast media, the renal parenchyma is evaluated with routine linear **nephrotomograms.** The final goal of the IVU is adequate distension and opacification of the **pelvocalyceal collecting system, ureters, and bladder** in order to visualize subtle abnormalities of the collecting ducts, urothelium, muscular wall, and retroperitoneal space.

B. Technique
1. **Preparation.** The optimal IVU begins with adequate patient preparation. Mild fluid restriction may be beneficial, but excessive dehydration may result in increased risk of contrast nephropathy. Patients undergoing intravenous urography should have **a mild laxative,** such as 1¼ oz (37.5 mL) of a standard extract of senna fruit, the evening before the examination. To avoid vomiting, most patients should have **"nothing by mouth after midnight."**
2. **Plain film evaluation.** The IVU begins with plain film evaluation of the kidneys, ureters, and bladder (KUB) and, if necessary, tomographic evaluation of the kidneys to evaluate for subtle stone disease. Exposure factors are determined so that optimal film quality can be obtained prior to the injection of intravenous contrast media.
3. **Nephrogram.** An adequate nephrogram is critical for the evaluation of renal size, masses, scars, and abnormalities involving renal parenchyma. The nephrogram has two components—vascular and parenchymal. The nephrogram is produced initially by the **vascular blush,** which is most prominent at 20–60 seconds after injection of intravenous contrast material, and by **tubular opacification,** which begins at 1–2 minutes postinjection. Linear tomograms of the kidneys (nephrotomograms) should be obtained between 1 and 4 minutes following the injection of intravenous contrast media to ensure maximum contrast enhancement of renal parenchyma.
4. **Pyelogram.** The quality of the pyelogram is determined by adequate opacification of urine in the renal calyces, infundibula, and renal pelvis and by good distension of the pelvocalyceal system. An adequate pyelogram is essential to visualize filling defects, obstruction, blunting of the collecting system, dilatation of the ducts of Bellini, and visualization of papillary cavities or strictures of infundibula. **Ureteral compression** is the essence of adequate pelvocalyceal visualization in most cases. Opacification of the collecting system is commonly obtained by intravenous administration of a total dose of 15 gm of iodine. An infusion (45 gm of iodine) or reinjection (23 gm of iodine) is required in some patients (*Radiology* 167:593–599, 1988). **Oblique radiographs are standard in some institutions,** though we rely heavily on the linear tomogram.

At approximately 8 minutes after the intravenous injection of contrast medium, a film of the kidneys is obtained with ureteral compression in place. Pelvocalyceal distension is usually maximal between 8 and 10 minutes. If distension is not adequate, a 10-minute film with ureteral compression can be obtained. If distension is still suboptimal, reinjection of 50 mL (23 gm I) of contrast media (i.e., Hypaque-90) is carried out, and ureteral compression is reapplied. Three- and five-minute films are subsequently obtained before compression is released.

5. **Ureters.** Visualization of the **proximal** ureters (opacification and distension) is usually obtained with the application of midureteral compression. A balloon device is inflated to compress the ureters over the sacrum. The **distal** ureters are usually visualized by obtaining a film of the abdomen and pelvis (KUB) immediately after the release of ureteral compression (usually at 10 minutes postinjection). If adequate visualization of the distal ureters is not obtained on this "release film," a **prone film** may demonstrate contrast media in the upper collecting system as well as in the distal ureters. Tomography of the ureters on a 14 × 17-inch film may aid in identifying the ureter in cases of decreased ureteral opacification (i.e., ureteral obstruction).

6. **Bladder. Adequate distension** of the bladder is necessary for a complete evaluation of the urinary tract. The 20-minute film of an IVU often adequately demonstrates anatomy and pathology of the bladder. If the bladder is not well distended, a delayed (30–40-minute) bladder film should be obtained. A **postvoid film of the bladder** as part of an IVU may be of help in evaluating bladder emptying, but it is not as accurate as urodynamic studies in the evaluation of bladder function. The mucosal pattern, visualization of diverticula, and detection of subtle mucosal filling defects may be visualized best on the postvoid film. Oblique views of the bladder also may be valuable in detecting periureteral diverticula and visualization of pathology involving the lumen, mucosa, bladder wall, or perivesicle space.

C. **Contrast media**
   1. **Physiology.** Intravenously administered contrast medium is excreted via the kidney by **glomerular filtration.** Tubular reabsorption and secretion do not play a signifi-

cant role in the physiology of renal handling of contrast media. Reabsorption of water and increased contrast media concentration are the only significant alterations of contrast material during tubular transit. Under normal conditions, salt and water reabsorption in the proximal tubule results in reabsorption of 80–90% of filtered water. This mechanism results in increased concentration of contrast media in the proximal tubule to approximately 5–10 times that of the plasma concentration. Only 10% of additional salt and water are reabsorbed from the distal tubules under the control of antidiuretic hormone (*Investigative Radiology* 5:473–486, 1970). The ultimate result is a very high concentration of contrast media within the renal tubules and collecting ducts. Hyperconcentration of contrast media in the tubules results in the dense **"tubular nephrogram"** that is optimal at approximately 2–4 minutes following the intravenous injection of contrast media.

The half-life ($T_{1/2}$) of contrast (time required for 50% of the administered contrast dose to be filtered by the kidney) is approximately 30–60 minutes in healthy adults (Fig. 5.1). The $T_{1/2}$ in elderly patients or those patients

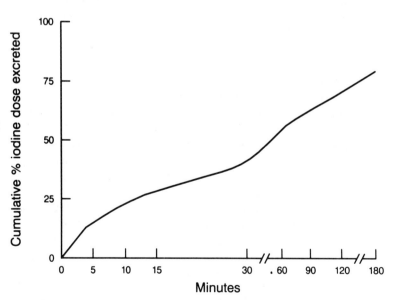

Figure 5.1. Cumulative percent of iodine dose excreted versus time (Hypaque 50%). (Modified from *Investigative Radiology* 17:494–500, 1982.)

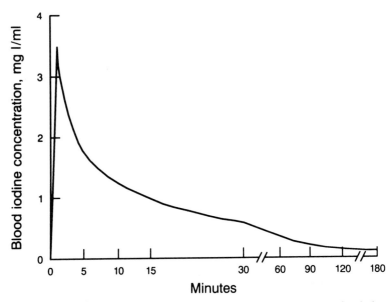

Figure 5.2. Blood iodine concentration (mg I per mL) versus time after bolus injection of 450 mg I per kg (Hypaque 50%). (Modified from *Investigative Radiology* 17:494–500, 1982.)

with impaired renal function is usually longer (60–180 minutes). After intravenous injection of contrast media, the contrast agent equilibrates between the serum and the extracellular fluid space ("third space"). Because a considerable amount of contrast ultimately lies within the extracellular fluid space, clearance of the contrast is dependent on the gradual diffusion of contrast media from the extracellular space (Fig. 5.2). Because of these factors, optimal imaging of the entire urinary tract is obtained in normal patients within the first 20 minutes after the injection of contrast media (Urography. In: *Radiographic Contrast Agents*. Gaithersburg, MD: Aspen Publishers, 1989, pp. 257–260).

2. **Chemistry.** Intravascular contrast medium is often conveniently divided into conventional high osmolar contrast media **(HOCM)** and low osmolar contrast media **(LOCM)** (Table 5.1). Conventional HOCM are all ionic and contain an anion (iothalamate or diatrizoate) and a cation (sodium, meglumine, or both). LOCM can either be ionic (ioxaglate) or nonionic (iopamidol, iohexol, and ioversol). HOCM containing sodium as their sole cation provide the

Table 5.1. Commonly Used Intravenous Contrast Media for Urography Rapid Injection Technique (IV Bolus)[a]

	Generic Name	Cation	Brand Name	Percentage in Solution	gm I per dL	Osmolality	Recommended Volume (mL) of Administration (Adult)
Ionic	Diatrizoate	Meglumine	Angiovist 282	60	28	1400	50 or 100
	Iothalamate	Meglumine	Conray-60	60	28	1400	50 or 100
	Diatrizoate	Meglumine	Hypaque 60%	60	28	1415	50 or 100
	Diatrizoate	Meglumine	Reno-M-60	60	28	1500	50 or 100
	Diatrizoate	Sodium 8% Meglumine 52%	Angiovist 292	60	29	1500	50 or 100
	Diatrizoate	Sodium 8% Meglumine 52%	MD-60	60	29	1539	50 or 100
	Diatrizoate	Sodium 8% Meglumine 52%	Renografin-60	60	29	1420	50 or 100
	Diatrizoate	Sodium	Hypaque 50%	50	30	1550	50 or 100
	Diatrizoate	Sodium	MD-50	50	30	1522	50 or 100
	Iodamide	Meglumine	Renovue-65	65	30	1558	50 or 100
	Diatrizoate	Sodium	Urovist Sodium 300	50	30	1550	50 or 100
	Diatrizoate	Sodium 29.1% Meglumine 28.5%	Renovist II	57.6	31	1517	50 or 100
	Iothalamate	Sodium	Conray-325	54.3	32	1700	50 or 100
	Diatrizoate	Sodium 30% Meglumine 60%	Hypaque-M, 90%	90	23	2938	50
Nonionic	Iopamidol		Isovue 300	61	30	616	50 or 100
	Iohexol		Omnipaque 300	64.7	30	709	50 or 100
	Ioversol		Optiray 320	68	32	702	50 or 100

[a] Adapted from H. W. Fischer. Catalog of Intravascular Contrast Media. *Radiology* 159:561–563, 1986.

highest urine concentrations. However, sodium causes more arm pain with injection and/or extravasation. HOCM containing meglumine as the sole cation appear to provide a slight improvement in distension of the collecting system but result in a lower urinary iodine concentration. Several contrast media combine sodium and meglumine salts. The clinical significance of these minor differences in cation characteristics is probably not significant.

LOCM are associated with less fluid shifts into the renal tubules after glomerular filtration. Therefore, the tubular concentration of LOCM is higher than that of the conventional HOCM. A denser nephrogram and slightly denser pyelogram is produced with LOCM (*Radiology* 162:9–14, 1987). However, the magnitude of the diuresis is less with LOCM and HOCM, and the degree of distension of the pelvocalyceal system is diminished. This diminished degree of distension with LOCM is particularly critical at lower does (50 mL, 300 mg I per mL solution) of contrast medium but is less critical at higher doses of contrast medium (100 mL, 300 mg I per mL solution). LOCM are lower risk but not **"no risk"** contrast agents.

3. **Concentration.** The optimal concentration (weight/volume) of contrast medium injected for intravenous urography is approximately 60%. The optimal concentration of contrast agents used for the infusion technique is approximately 30% (Table 5.2). The lower concentration utilized in the infusion technique allows for the concomitant administration of a larger amount of fluid to be administered with the iodinated contrast material. This additional fluid is available for diuresis and can aid in optimally distending the pyelocalyceal collecting system whether or not ureteral compression is used for the examination.

4. **Viscosity.** The viscosity of most conventional ionic agents (i.e., Conray-60, 6.0 cp at 25°C) permits injection of these agents at room temperature without difficulty. Some ionic agents (Hypaque-90) utilized for reinjection purposes do have a much higher viscosity (34.7 cp at 25°C). In addition, several of the newer nonionic agents also have a slightly higher viscosity (i.e., Omnipaque 300, 10.35 cp at 25°C) than the conventional ionic agents (New Developments in Radiographic Contrast Media. In: *Contrast*

Table 5.2. Commonly Used Intravenous Contrast Media for Urography Infusion Technique (Rapid IV Drip)

	Generic Name	Cation	Brand Name	Percentage in Solution	gm I per dL	Osmolality	Total gm I in 300 mL	Recommended Volume (mL) of Administration (Adult)
Ionic	Iothalamate	Meglumine	Conray-30	30	141	600	42	300
	Diatrizoate	Meglumine	Hypaque 30%	30	141	633	42	300
	Diatrizoate	Meglumine	Reno-M-Dip	30	141	566	42	300
	Diatrizoate	Meglumine	Urovist	30	141	640	42	300
	Diatrizoate	Meglumine	Hypaque 25%	25	150	696	45	300
Nonionic	Iohexol		Omnipaque 250	51.8	30	504	48	200
	Ioversol		Optiray 320	68	32	702	48	150

Media: Biologic Effects and Clinical Applications. Boca Raton, FL: CRC Press, 1987, Vol. 1, pp.28–44). These agents with higher viscosity often require heating to body temperature (37°C) for easy intravascular administration.

5. **Dose and volume.** Maximum iodine concentration in the urine is obtained with ionic contrast media utilizing a dose of 300 mg I per kg body weight (*AJR* 113:423–426, 1971). It is our belief that the risk of contrast nephropathy increases with an increasing dose of contrast agent. Because of this and other risk factors, it is prudent to select the lowest dose of contrast necessary to perform an adequate examination. In our practice, 50 mL (300 mg I per mL), with good ureteral compression, is adequate for routine use in urography in most adult patients. If distension of the collecting system is inadequate after the rapid injection of 50 mL of intravenous contrast media (approximately 10% of patients), we commonly reinject with 50 mL of Hypaque-90 (23 gm I). This combination results in a total dose to the patient of 38 gm I (15 and 23 gm I). The decision to reinject is made if distension of the collecting system is not adequate on the postinjection 10-minute film.

Infusion of a larger amount of contrast is recommended in some circumstances (Table 5.3). The infusion technique is useful in **obese patients** when adequate ureteral compression cannot be achieved, in patients with an **abdominal aortic aneurysm,** in patients who have had **recent abdominal surgery, and in patients with acute or chronic ureteral obstruction.** With the infusion technique, a total dose of 42 gm I is delivered (300 mL Reno-M-Dip). Therefore, in our practice, we rarely exceed 50 gm I of intravenously administered contrast media at any one setting.

Table 5.3. Situations That May Warrant Infusion Technique in Intravenous Urography

---

Large patients (>200–250 pounds)
Recent abdominal surgery
Abdominal aortic aneurysm
Renal insufficiency (serum Cr > 2 mg/dL)
Ureteral diversion/cystectomy
Acute stone disease
Lasix IVU (flow-dependent obstructions)

Patients who weigh **over 200–250 pounds** may distribute more contrast into a larger extracellular third space, and thus, less contrast is available for renal excretion. A higher dose of contrast media may be administered by infusion technique in these patients. Infusion of the larger amounts of contrast media is also useful in patients with **compromised renal function** to ensure adequate visualization of the renal parenchyma and collecting system. Finally, the infusion technique of a larger amount of contrast is also of help in patients with **acute stone disease** in which the diuretic effect of larger amounts of contrast media may actually be therapeutic. The advantages and disadvantages of the infusion technique are listed in Table 5.4.

6. **Route of administration.** The route of administration of intravenous contrast media is generally through an **antecubital vein.** A rapid bolus, hand injection of intravenous contrast media is done routinely. Generally, an 18-gauge needle is used for the injection of contrast media to allow for rapid adminstration. When the **infusion technique (300 mL)** is used, rapid drip infusion by gravity is done through an 18-gauge needle, preferably into the antecubital vein. Rapid infusion using this technique is usually completed within 3–5 minutes.

Use of a smaller gauge needle or a smaller vein may impair the rapid injection or the rapid drip infusion of the contrast media. Use of a smaller vein (i.e., dorsum of the hand) may not accommodate the volume of rapidly injected contrast media, and rupture of the vein and extravasation of the contrast media may occur. If veins on the dorsum of the hand are employed, the injection rate should be slower to reduce the risk of vein rupture. If extravasation occurs, it is our practice to **elevate the affected limb and apply heat immediately.** The patient is observed for 1–2 hours in the radiology department, and

Table 5.4. Patients at Risk for Contrast Neuropathy

Definite risk	Diabetics with renal insufficiency
Possible risk	Any patient with renal insufficiency
	Congestive heart failure
	Concomitant use of nephrotoxic drugs
	Multiple myeloma
	Closely spaced contrast media examinations

the patient's primary physician is notified to ensure close observation and follow-up.

The rate of administration of contrast media is important in terms of optimal filming of the nephrogram and pyelogram. Rapid injection provides an optimal dense nephrogram at approximately 2 minutes and enables adequate evaluation of the renal parenchyma. Although rapid injections are well tolerated, they may be associated with a higher incidence of flushing, nausea, and vomiting. Rapid injections of contrast media are generally not associated with an increase in severe reactions.

D. Special conditions
1. **Renal insufficiency.** Ultrasound, CT, and retrograde pyelography are often employed in the imaging of the genitourinary tract in patients with renal failure. Deterioration of renal function is a major concern in patients with renal insufficiency who are undergoing an examination requiring intravenous contrast media. Patients with diabetes mellitus and associated renal insufficiency have an increased risk of progressive deterioration of renal function. Therefore, caution should be exercised in performing intravenous urography or CT with contrast enhancement in diabetic patients who have renal insufficiency (serum creatinine > 1.5 mg/dL) (*New England Journal of Medicine* 320:149–153, 1989). Other patients who might also be at risk for developing contrast nephropathy are listed in Table 5.5. Intravenous urography can be performed in these patients if clinically warranted and the patient has been adequately hydrated prior to the examination. Adequate hydration prior to intravenous contrast administration is an important factor in reducing the risk of contrast nephropathy in these patients. Some researchers believe a fluid challenge (250–500 mL of 0.9% NaCl solution) or a small dose of a diuretic (Lasix, 20 mg

Table 5.5. Situations That May Warrant the Use of LOCM in Intravenous Urography

History of significant prior contrast reaction
History of asthma or multiple allergies
Severe heart disease
Debilitated patients
Avoidance of motion artifact
Extremely anxious patients

IV) may be of help in "flushing" the iodinated contrast material out of the kidney to lessen the likelihood of potential contrast nephropathy. Intravenous urography provides little if any information in patients with a serum creatinine > 4 mg/dL and is not recommended in these patients.
2. **Hypertensive intravenous urogram.** Although the sensitivity (60%) of hypertensive intravenous urography is relatively low, the examination continues to be utilized in selective hypertensive patients who have been referred for intravenous urography for other indications (*AJR* 138:43–45, 1982). To detect a subtle **delay in appearance time** (appearance of contrast media in the collecting system), contrast media must be injected via a bolus, and 2- and 3-minute films of the kidneys must be precisely timed. The 2-minute film should be obtained exactly 2 minutes from the **start** of the injection or infusion. The 3-minute film should be obtained in a similar fashion. The recommended dose and mode of injection in hypertensive IVUs is rapid injection of 50 mL of contrast media (15 gm I). The timing of the 2- and 3-minute films following the injection of contrast media is critical. By adhering to these strict techniques, the sensitivity of the hypertensive IVU can be maximized.

In addition to a subtle delay in appearance time, other positive findings during a hypertensive IVU are: discrepancy in renal size, ureteral notching from vascular collaterals, hyperconcentration of contrast media in the collecting system of the affected kidney, and a suboptimally distended ("spidery") pelvocalyceal system.
3. **Postoperative patients.** Patients who have recently undergone **abdominal surgery** often cannot tolerate ureteral compression. In these patients, the infusion technique should be utilized to maximize pyelocalyceal opacification and distension. In addition, those patients who have had cystectomy and ureteral diversion (i.e., **ilioconduit**) should have an infusion of contrast media, since adequate ureteral compression may be difficult.
4. **Lasix intravenous urogram.** In most cases, ureteropelvic junction or ureterovesical junction obstructions are readily apparent on the IVUs. However, when the diagnosis is uncertain or volume-dependent obstruction is suspected, a Lasix IVU may be of help. The concept is based on the

premise that a large urine flow rate, after Lasix administration, may precipitate flank pain and flow-dependent obstruction. It is best to utilize an infusion of contrast media and to administer 20 mg of Lasix IV after the 10-minute film. The precipitation of flank pain, increased distension of the renal pelvis, and delayed washout of contrast from the renal pelvis are all positive signs that suggest flow-dependent obstruction.

II. **Cystography.** The goals of cystography include proper distention and opacification of the bladder along with multiple views of the bladder in order to detect a variety of abnormalities, such as filling defects, bladder wall pathology, diverticula, and reflux. Detection of vesicoureteral reflux is most important in children but is also an important component of the examination in many adults.

Cystography is the standard examination for bladder trauma, and careful attention to technique is vital in order to detect bladder injury. Additional indications for cystography include bladder dysfunction, persistent hematuria, prolapsing urethral mass, megalourethra, large bladder, prune belly syndrome, and lower urinary tract anomalies, such as hypospadias or epispadias.

A. **Technique.** Cystography is usually performed after the transurethral placement of a Foley catheter. A scout plain film of the bladder is obtained to exclude calcification in the wall of the bladder (i.e., schistosomiasis), gas in the lumen or wall of the bladder (i.e., emphysematous cystitis), and bladder calculi. Contrast media are introduced through the catheter into the bladder via gravity drip with the bottle approximately **1 meter** above the table top. This technique permits maximal distension of the bladder without the risk of forceful overdistension and possible bladder rupture. Fluoroscopic observation of the bladder during filling is performed in addition to anteroposterior (AP), oblique, lateral, and postvoid spot films. Postvoid films of the bladder are an essential part of the examination.

Patients with abdominal trauma often require a unique approach to cystography. In the event of a pelvic fracture, a retrograde urethrogram should always be performed prior to the cystogram to prevent blind passage of a urethral catheter through a urethral disruption or a partial urethral injury. If a **urethral injury** is identified, a cystogram should still be performed. This can be accomplished after fluoroscopic or cystoscopic placement of a Foley catheter into the bladder. If transurethral entry is not possible, percutaneous suprapubic catheterization can be used (*Journal of Pediatrics* 81:555–558, 1972). In all cases of trauma, an AP, right posterior oblique, left posterior oblique, and lateral film should be obtained. A postvoid film is also critical, since it may be the only film demonstrating extravasation of contrast.

Bladder tears can either be **extraperitoneal** (most common) or **intraperitoneal**. Because intraperitoneal rupture can be difficult to detect on cystography, fluoroscopy of the paracolic gutters and a postvoid film that includes the whole abdomen should be obtained. If intraperitoneal bladder tear is strongly suspected, postcystogram CT of the abdomen and pelvis may be required. In addition, the use of **denser contrast medium** (30–60% (weight/volume)) should be considered to demonstrate small tears more accurately.

B. **Contrast media.** Meglumine salts of diatrizoate and/or iothalamate are the most widely used contrast media for cystography. Meglumine salts are less irritating to bladder mucosa.

Solutions containing 15% (weight/volume) contrast media are generally optimal for cystography. Because the bladder capacity is large, solutions containing greater than 15% (weight/volume) contrast media result in very dense opacification of the bladder. This density could preclude identification of subtle abnormalities in the bladder. In addition, higher concentrations of contrast media (i.e., 30% (weight/volume)) may be irritating to the bladder mucosa (*Radiology* 104:563–565, 1972). Similar contrast media can be utilized in pediatric cystography.

Absorption of contrast media from the bladder during a cystogram may occur. Therefore, patients who have had a severe reaction to intravenous contrast media should be evaluated carefully prior to the examination. LOCM (200 mg I per mL) may be indicated in these patients. Nuclear scintigraphy, CT without IV contrast, MR imaging, or ultrasound can also be utilized in selected patients if contrast cystography is contraindicated.

C. **Voiding cystourethrography.** Fluoroscopically monitored voiding cystourethrography is performed in order to evaluate the bladder adequately, to exclude the possibility of vesicoureteral reflux, and to provide adequate visualization of the urethra. Because vesicoureteral reflux is more likely to occur with greater intraluminal bladder pressures, fluoroscopy should be performed to detect reflux during voiding.

If vesicoureteral reflex occurs, it is important to document the level of reflux (i.e., distal ureter, intrarenal collecting system, etc.). (See also Chapter 20, "Pediatric Contrast Agents.")

It is also important to evaluate the urethra fluoroscopically and with "spot" films. Spot films are generally taken of the urethra in the AP, oblique, and lateral projections. The size and caliber of the urethra are important in order to exclude focal strictures (i.e., status posturethritis) and/or dilatation (i.e., posterior urethral valves). The posterior urethra is best visualized during voiding cystourethrography.

Since urethral volume is small, dense opacification of the urethra is required. A reasonable compromise to provide adequate opacification of the urethra and of the bladder is to use a 30% (weight/volume) contrast media. In the pediatric population, some have advocated a lower concentration of contrast media (i.e., 15–20% (weight/volume)) during voiding cystourethrography in order to decrease the likelihood of irritation of the bladder mucosa.

D. **Complications. Bladder infections** can occur in up to 6% of patients after cystography (*Annales de Radiologie (Paris)* 13:283–287, 1970). Therefore, cystography in patients with active infections should be avoided. In tetraplegic patients (lesions above T-5), filling of the bladder and forcible opening of the bladder neck may cause a severe reaction including severe headache, sweating, hypertension, and bradycardia. An alpha-adrenergic blocker prior to the cystogram can help prevent this **autonomic dysreflexia** (*Seminars in Roentgenology* 18:255–256, 1983).

III. **Urethrography**
  A. **Goals.** The goals of urethrography are adequate opacification and distension of the urethra and visualization of urethral anatomy. Urethrography is employed in the evaluation of patients with possible urethral trauma, urethral strictures, urethral diverticula, and fistulous communications.
  B. **Technique.** In males, retrograde urethrography is usually performed by inserting a nonlubricated tip of an adequately sized catheter (18 French for adults) into the fossa navicularis. Some type of clamping mechanism is often used to keep the tip of the catheter in the distal urethra. Sometimes, it is easier for the radiologist to compress the glans of the penis as a mechanism of keeping the tip of the catheter in place during the examination. Care must be taken to keep the

physicians hand out of the fluoroscopic field to minimize radiation exposure. Films are taken in both oblique positions under fluoroscopic guidance. Sufficient pressure (i.e., hand injection) is required to overcome the resistance of the external sphincter. Adequate distension and visualization of the bulbous and pendulous portions of the penile urethra are usually accomplished. The membranous and prostatic portions of the urethra are often not adequately visualized on retrograde urethrography because of the inability to distend the urethra adequately as a result of high-resistance pressure at the membranous urethra. If visualization of the prostatic and membranous urethra is desired, a catheter can be placed into the bladder to initiate an antegrade voiding cystourethrogram.

Retrograde urethrography is not routinely utilized in female patients because of the short length of the female urethra. A double-balloon catheter can be used to study the female urethra and to exclude the possibility of **urethral diverticula** (*AJR* 136:259–264, 1981). Recently, MR imaging and transvaginal ultrasound of the urethra have been shown to be sensitive means of detecting urethral diverticula.

C. **Contrast media.** Contrast media for urethrography must be sufficiently viscous and dense for proper visualization of the urethra. It appears that meglumine salts of diatrizoate or iothalamate are most useful for the examination. If intravasation occurs, the meglumine salts appear to be less painful and irritating than sodium salts. Systemic reactions following retrograde urethrography are extremely rare even in patients with intravasation.

Generally, 60% (weight/volume) solutions are used in retrograde urethrography to provide maximal opacification of the urethra. Small fistulous tracts communicating with the urethra, subtle urethral strictures, or small areas of extravasation can be seen with proper technique. In children, 30% (weight/volume) contrast media solutions should be used.

D. **Complications.** Because contrast material is injected under pressure during retrograde urethrography, intravasation of contrast material into the corpus spongiosum may occur. This most commonly occurs in patients with urethral strictures and/or urinary tract infections. Antibiotic coverage is recommended following the procedure if intravasation has occurred (*Journal of Urology* 112:608–609, 1974).

Chapter / 6
Contrast Media Use in Computed Tomography

Richard A. Leder
N. Reed Dunnick

Intravascular contrast material is an essential tool for the performance of abdominal and pelvic computed tomography (CT). The ability to characterize pathology in the liver, spleen, pancreas, and kidneys is aided by the use of intravenous contrast media. Retroperitoneal structures are better defined by enhancement of the aorta and inferior vena cava, and the detection of iliac lymphadenopathy is improved by enhancing the pelvic vessels with iodinated contrast media. Delineation of disease in the chest including the mediastinum and hilar regions is best accomplished when intravenous contrast material is used.

While the utility of intravenous contrast material is not in question, there is no consensus as to the best method of injecting contrast or of the optimal scan protocols. Additionally, there are many contrast agents from which to choose. Thus, each examination must be tailored to the individual patient, and specific clinical question posed. Bowel opacification is discussed in Chapter 18.

I. **General indications. Contrast-enhanced** abdominal CT reveals the relative vascularity and vascular characteristics of a mass. Contrast enhancement is of help in discriminating vessels from neoplastic masses, maximizes anatomic and lesion detectability, and opacifies the urinary tract with excreted contrast material. Subtle or occult lesions may only be detectable on contrast-enhanced images. Since normal and abnormal tissue handle intravascular contrast material differently, the attenuation value difference usually increases after the contrast administration. Since intravenous contrast material increases the sensitivity of CT examinations, it is used in the majority of cases.

In the search for intra-abdominal or intrathoracic pathology, there are only a few circumstances in which intravenous contrast impairs disease detection and characterization. One example would be the patient with a hypervascular liver metastasis that might become isodense upon the administration of intravenous contrast. The lesion would then be difficult or impossible to detect. A thorough search should be performed, both of the patient's chart and of any prior radiographs to determine those situations when intravenous contrast either should not be given or should be given only after initial noncontrast imaging.

There are many circumstances in which disease can be detected without the use of intravenous contrast. However, more information can be gained by providing contrast enhancement. This would even include the scan performed to rule out an intra-abdominal or intrathoracic source of infection. Questionable intra-abdominal fluid collections that are nonspecific on noncontrast scans can be established to be urinomas when they opacify with excreted contrast material. Pleural enhancement in the chest aids in the diagnosis of an empyema and the distinction from a pulmonary abscess.

A. **An advantage** of the routine administration of contrast material is a decrease in the time required to perform a study. Noncontrast images are seldom performed, normal and anomalous vascular structures are immediately identified, and a thorough search for disease has been accomplished.

B. **Disadvantages** of the routine use of contrast material include those patients with a disease process that can be diagnosed without intravenous contrast material. They will have been unnecessarily subjected to intravenous contrast material with all its attendant risks. However, these risks have greatly diminished with the use of nonionic contrast agents, and the potential gains from the routine use of intravenous contrast agents, and the potential gains from the routine use of intravenous contrast outweigh the disadvantages.

II. **Specific indications.** Most CT scans are performed to answer specific clinical questions, usually the presence or absence of disease in a specific organ. A large percentage of CT scans are performed to assess the liver. Scanning is frequently performed to evaluate potential pancreatic masses, splenic lesions, renal disease, retroperitoneal lymphadenopathy, adrenal masses, abdominal aortic aneurysms, and pelvic masses or adenopathy, as well as to detect a source of intra-abdominal infection. The greatest need for intravenous contrast in the chest lies with defining mediastinal structures to exclude mediastinal masses and evaluate potential hilar adenopathy. Contrast also aids in the evaluation of disease in the pleural space. The individual needs of each organ system must be addressed prior to determining the individual scan protocol best suited to answer the clinical question asked.

III. **Techniques of contrast administration.** Once it has been decided to administer intravenous contrast material, a decision must be made about how best to inject it. The aim is to distinguish between lesions and normal tissue as well as to differentiate vascular structures from adjacent lymph nodes. For the purpose of this discussion, the effect of the contrast injection may be classified into one of three categories. The **bolus phase,** obtained immediately after a bolus injection of intravascular contrast material, is defined as an attenuation difference of 30 or more HU between the aorta and the inferior vena cava. The **nonequilibrium phase** is seen after the bolus effect and is defined as an aorta-caval difference of 10–30 HU. Subsequently, an **equilibrium phase** is achieved where the attenuation difference from the aorta to the inferior vena cava is less than 10 HU. The contrast administration options include: drip infusion, bolus technique, dynamic sequential bolus CT, CT portography, and hepatic CT angiography.

   A. **Drip infusion.** The drip infusion technique is characterized by its ease of performance. Contrast material is administered into a superficial arm vein over a several minute period (approximately 5–10 minutes). Scanning begins after the initiation of the contrast material infusion. Compared with precontrast examinations, visualization of hepatic tumors is improved in the nonequilibrium phase but not in the equilibrium phase. The equilibrium phase is reached 2 minutes after a contrast material bolus and with termination of an infusion. Scanning in the equilibrium phase does not improve visualization of tumors in the liver when compared with a precontrast examination and carries a considerable risk of tumor enhancement. **This scanning technique is strongly discouraged** (*American Journal of Roentgenology* 140:291–295, 1983).

   B. **Bolus technique.** The bolus technique requires scanning immediately following a rapid bolus injection of contrast. A volume of 50–100 mL of contrast is administered in a short time period (approximately 10–20 seconds). Scanning must begin **immediately** after the termination of the bolus injection such that images are acquired prior to the acquisition of the equilibrium phase. With earlier scanners possessing only a modest tube heat capacity, this technique had limited usefulness, as repeat boluses were required to evaluate the entire liver. Compared with precontrast examinations, there is a significant improvement in the visualization of hepatic tumors when images are obtained by using this technique before the equilibrium phase.

C. **Dynamic sequential bolus CT.** With this technique, multiple separate CT slices can be obtained each minute, thereby allowing evaluation of the entire liver or of a section of the abdomen in a short period of time. Hepatic CT usually requires from 12 to 20 contiguous 10-mm sections (average, 16) and can be achieved within 2 minutes after the initiation of scanning. Contrast material is injected in a biphasic mode, from either a power injector or a strong hand bolus. The bolus contrast-enhanced dynamic scan is reproducible and is useful for patients undergoing sequential CT for assessment of tumor bulk in response to chemotherapy. This technique can also be employed in scanning other parts of the body, including pancreatic lesions, and for staging lung carcinoma (*Radiology* 170:617–622, 1989).

D. **CT portography.** The liver receives 25% of its supply from the hepatic artery and 75% from the portal system. However, hepatic neoplasms derive nearly all of their blood supply from the hepatic artery. Contrast reaching the liver is delayed for several seconds, as it must pass through the bowel to reach the portal vein. Hepatic metastases will have a greater local iodine concentration and result in hyperattenuation on CT scans obtained after selective hepatic artery injection. Arterial portography allows contrast to be delivered selectively into the portal venous system without distribution to and dilution with the central blood volume. This will result in greater hepatic parenchymal enhancement and increased contrast differentiation between focal lesions and the background. This technique is easier to implement than hepatic artery injection CT, in that the catheter tip needs only to be placed in the superior mesenteric artery distal to any anomalous hepatic artery branches.

E. **Hepatic CT angiography.** With this technique, femoral artery catheterization is performed and a catheter is placed into the celiac artery (*Radiology* 159:685–691, 1986). Hepatic angiography and CT hepatic angiography are used to determine tumor resectability. This is an invasive technique. However, it does allow for improved preoperative evaluation of candidates for liver resection to define vascular anatomy and to determine the extent of tumor.

Presently, the favored technique for CT scanning is dynamic sequential bolus CT. Current generation scanners are capable of performing this technique, and with mechanical pump injectors, this technique currently provides the best scan quality.

IV. **Contraindications to the administration of intravenous contrast material.** An individualized list of contraindications is likely to exist at each institution. The following contraindications to intravenous contrast material (some of which are listed in Table 6.1) are commonly included.

   A. **Prior major reaction to contrast media.** Alternative studies can frequently be performed such that intravenous contrast material need not always be given to document disease. Alternative studies include ultrasound and magnetic resonance (MR) imaging. Given the combination of ultrasound and MR, virtually all intra-abdominal organs, the retroperitoneum and the pelvis can be evaluated safely without intravenous contrast. Furthermore, disease may be documented on a noncontrast CT examination.

   B. **Renal failure.** Since preexisting azotemia is the most common risk factor for contrast-induced renal failure, intravascular contrast material should be avoided or minimized in these patients (see Chapters 3 and 4). This does not apply, however, to patients undergoing dialysis. It is prudent to discuss with the referring physicians the timing of the contrast injection relative to dialysis.

   C. **Known or suspected pheochromocytoma.** It is our practice not to administer contrast material to patients with pheochromocytoma for fear of inciting a hypertensive crisis. Alternative techniques such as MR imaging or radionuclide examination with metaiodobenzylguanidine (MIBG) are available to document disease in these patients. Additionally, we do not administer contrast material to patients with

Table 6.1. Contraindications to Intravenous Contrast

Major reaction to contrast media
Renal failure (creatinine over 3.0 mg) unless predialysis
Pheochromocytoma/myasthenia gravis
Hypokalemia/hyperkalemia

---

myasthenia gravis, as there is a small risk of inducing myasthenic crisis (respiratory arrest) when iodinated contrast materials are given.
D. **Abnormal serum potassium.** In patients with a markedly abnormal serum potassium, hypokalemia or hyperkalemia, we are reluctant to give intravenous contrast material. Electrolyte changes after the administration of contrast material may cause cardiac arrhythmias, and therefore, we advise correction of the patient's potassium imbalance prior to the administration of contrast material. Hypokalemia is more arrhythmogenic than hyperkalemia.
E. **Contrast material can be administered to patients who have sickle cell anemia who are not in active crisis.** Similarly, contrast can be given to patients with **multiple myeloma** who are not in renal failure and do not have proteinuria. The literature on these select patient groups is not clear, and the examination must be tailored to answer particular clinical questions. Frequently, this can be done without intravenous contrast or by using an alternative examination.

V. **Types of contrast.** A wide array of intravascular contrast media are available for CT. These agents differ in their **ionicity, iodine content, percent concentration, and osmolality.** Prior to the introduction of nonionic contrast media, most agents were high osmolar ionic material. Typically, these agents had osmolalities greater than 1000 mOsm/kg (well above serum osmolality), and their injection was frequently complicated by nausea and vomiting. Other more serious side effects include potential anaphylactic reactions. With the introduction of nonionic contrast agents with significantly lower osmolality, reactions greatly decreased. It is our current practice to give nonionic contrast media to all patients undergoing body CT.
A. **Ionic versus nonionic.** Since features have been identified to place patients in a higher risk group for an adverse contrast reaction, many institutions as well as the Committee on Drugs and Contrast Media of the American College of Radiology (ACR) have developed guidelines for the selective use of nonionic contrast media. Those patients not included in the high-risk group are at very low risk for adverse reaction to intravascular contrast media and receive conventional ionic agents. The groups initially identified by the ACR to receive nonionic contrast material included: (1) patients with a **previous significant adverse reaction** to contrast material, a **strongly allergic history,** or asthma; (2)

patients with **cardiac dysfunction** including severe arrhythmias, unstable angina pectoris, recent myocardial infarction, pulmonary hypertension, and congestive heart failure; (3) patients with **generalized severe debilitation;** (4) patients undergoing **potentially painful examinations** such as peripheral arteriography, external carotid arteriography, and lower limb phlebography; and (5) patients undergoing examinations such as **digital angiography where inadvertent motion must be minimized** in order to optimize image quality (*American College of Radiology Bulletin* pp. 9–10, November 1988).

This approach has limitations. There are no scientific or clinical criteria to identify all patients who will have an adverse contrast reaction. "Patient-specific indications" are considered unreliable and cannot replace clinical judgment. "Procedure-specific indications" have not been clearly identified. A review of the recommended guidelines for identifying high-risk patients and procedures would include the vast majority of patients being examined. This proved to be true in our experience (*Investigative Radiology* 26:17–21, 1991). Cost containment based on these guidelines would not have a significant economic impact and would not justify either the risk or the expense and effort involved in attempting to apply them (*Current Problems in Diagnostic Radiology* XX:49–88, 1991).

Improved patient tolerance of nonionic material is important, particularly in light of the necessity to employ a dynamic sequential bolus CT technique. Patient movement during an examination can cause delays, discontinuation of the study, or suboptimal image quality. This is particularly important in the CT evaluation of the liver, pancreas, and kidneys. Pain or nausea caused by contrast material will almost certainly lead to excessive motion. Tolerance is a particularly important factor in pediatric patients, as vomiting in the sedated pediatric patient carries an increased risk of aspiration. Patient perception of heat and pain are lower with low osmolar contrast media, and therefore, there are fewer mild to moderate side effects with these agents. The results of the Royal Australasian College of Radiology (RACR) intravenous contrast media (IVCM) survey as well as the Japanese IVCM survey substantiate the use of routine nonionic contrast material based on its broad safety profile.

The excellent work performed by the RACR survey

supervised by Dr. John Palmer and the work by Dr. Hitoshi Katayama have given radiologists **solid evidence of the safety of nonionic contrast media compared with ionic contrast material.** In the Japanese study, a severe adverse reaction occurred in only 0.04% of patients after the injection of nonionic contrast material. In the report of the RACR study, only 0.02% of patients experienced a serious adverse effect with nonionic contrast material, and no deaths were reported in this group. Furthermore, results from both studies indicated that a patient was statistically safer if he or she were a high-risk patient receiving nonionic contrast material than a low-risk patient receiving high osmolar contrast material. It appears that patients are safer by a factor of approximately 6 when they receive nonionic contrast material (*Australasian Radiology* 32:426–428, 1988; *Radiology* 175:621–628, 1990).

B. **Extravasation.** Our bias toward nonionic contrast materials is based not only on the systemic safety profile of these agents but also on potential local complications from the bolus injection of contrast material. It has been well established that the best and most reproducible technique for contrast administration is that delivered through a power injector with scanning performed in dynamic mode. If contrast extravasation occurs, a larger amount will be injected outside the vein than if a hand injection technique is used. The time required for the physician/nurse/technologist to recognize the extravasation and to react by interrupting the injection will be far greater when a mechanical pump injector is used than when the contrast material is injected by hand. It is our current practice to use a pressure-limited low-rate injector for the injection of contrast material in abdominal and thoracic CT. However, extravasations are likely to be more common and more severe with the routine use of the power injector. A recent report found that extravasations occurred in 6 patients after 5280 injections using the power injector for CT (*Radiology* 76:69–70, 1991). This represents a complication rate of 0.1%. No long-term sequelae resulted from these extravasation injuries. Research in laboratory animals has demonstrated that **nonionic contrast media are less injurious to cutaneous/subcutaneous tissue than conventional high osmolar contrast material** (*Investigative Radiology* 25:504–510, 1990). The injuries caused by ionic contrast agents in rats include severe necrosis, edema, and hemorrhage. It is our clinical experience that in the few patients that have sustained contrast extravasations, all with nonionic agents, there is no significant morbidity associated with the extravasation even when the volume of contrast material is large. Furthermore, we have seen such an improved tolerance to nonionic contrast that it is extremely rare to have to interrupt a dynamic CT examination because of patient pain, nausea, vomiting, or even contrast reaction.

VI. **Available contrast media.** The list of available iodinated intravenous contrast media is long. Among the various agents, the three characteristics that are important in regard to their use in CT are **iodine content, viscosity, and osmolality.** These factors will be of help in determining the ability to inject quickly a large amount of iodine in a small volume and influence the number and extent of adverse reactions. The features of the commonly used agents are listed in Table 6.2. We exclusively use nonionic agents, presently iopamidol 300 and ioversol 320.

VII. **Dosage.** It is unclear at this time whether there is an optimal dosage for intravenous contrast material. The volume of contrast injected should be individualized to the particular patient and the examination or examinations being requested. For example, if the liver is the only area of clinical interest, it is possible that less contrast may be given to scan this organ alone. If it is necessary to stage tumor burden in the liver, assess retroperitoneal adenopathy, and determine pelvic adenopathy, it is likely that a greater amount of contrast must be given. Furthermore, if tumor burden in the chest is also part of the

Table 6.2. Physical Factors of Current CT Contrast Media

		% Concentration	Total Iodine (mg/mL)	Osmolality (mOsm/kg)	Viscosity (cp @ 37°C)
Ionic	Renografin-60 (Squibb)	60	288	1420	4.0
	Conray-60 (Mallinckrodt)	60	282	1440	3.8
	Hypaque 60% (Winthrop)	60	282	1340	3.8
Nonionic	Iopamidol 300 (Squibb)	61	300	616	4.7
	Ioversol 320 (Mallinckrodt)	68	320	702	5.8

study, the possibility of either a separate bolus in the chest or a large amount of contrast may be necessary.

It is our current practice to inject approximately 150 mL of iopamidol 300 in an average-size adult patient. In adult patients weighing approximately 100 pounds, the amount of contrast is limited to 100 mL. If the patient weighs less than 100 pounds, we use approximately 2 mL/kg. This is the same dose given to pediatric patients. It is certainly possible that with newer contrast agents possessing different concentrations of iodine, the total volume of contrast may successfully be decreased. Given the cost of the newer nonionic agents, the desire to limit contrast volume is always a strong consideration.

II. Route

A. Routine injections.
Only proximal forearm or medially placed antecubital veins should be used as sites for the administration of intravenous contrast. This is especially important for dynamic studies that require a bolus effect. In certain rare instances, peripheral forearm and hand veins may be used. When the automated power injector is used, 2-inch, 18- or 20-gauge Quik-Caths must be used. If a shorter or smaller gauge Quik-Cath is the only possible means for IV access, then a hand injection should be performed.

B. Infusaport/Portacath.
Infusaports and Portacaths are not used for the bolus injection of intravascular contrast. They may be used in specific cases when nondynamic studies are being performed. For a patient with an Infusaport or Portacath, a peripheral IV should be started and used for the contrast administration. If an adequate peripheral IV line cannot be started, then the scan may be performed without IV contrast. It is the responsibility of the physician or nurse administering contrast material to flush these catheters adequately with saline before and after the administration of contrast.

C. Hickman catheter.
Hickman catheters can also be used for the administration of IV contrast. Only hand injections may be performed. If resistance is encountered, a peripheral line can be started. These catheters should be flushed with saline before and after contrast administration.

III. Scan protocols.
Scanning protocols are constantly changing. Factors influencing these constant changes are the tube heat capacity and software updates in current generation scanners. The aim is to perform scans as rapidly as possible such that contrast levels in organs and vessels are peaking during

scanning. Software improvements allow a greater number of slices to be obtained with shorter interscan delays. Solid scintillation detectors have improved geometric absorption and scintillation efficiency, less afterglow, and improved stability when compared with xenon detectors. Smaller apertures improve spatial resolution (*Clinical Imaging* 13:189–194, 1989).

The peak of tissue enhancement represents the period of greatest lesion detectability within intra-abdominal organs. The attenuation of liver parenchyma, for example, should be abruptly elevated by 60–80 HU and maintained in that range during the scanning period. Liver enhancement can be held above 50% of peak value for 80–100 seconds with contrast volumes of 60–80 mL. Pancreatic tissue enhancement is short lived, with 50% of peak value maintained for only 40–60 seconds. A delay between the initiation of a contrast bolus and the initiation of scanning allows time for peak enhancement, of both organs and vessels (Fig. 6.1). Timing is particularly important in the liver to enhance both portal vessels and hepatic veins.

Injection speed and volume are both important factors in maximizing contrast enhancement. Arteriovenous iodine differences can be measured on CT images and used as a means of assessing the vascular opacification obtained by a given technique. Three phases of intravascular enhancement exist: the **bolus effect phase,** the **nonequilibrium phase,** and the **equilibrium phase.** Maximal vascular enhancement (the bolus effect)

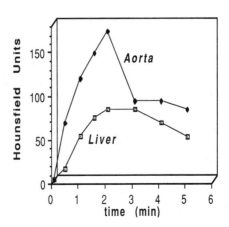

Figure 6.1. Effect of bolus injection of contrast material. Aortic and vascular CT attenuation peaks followed by a biphasic decrease. Hepatic enhancement plateaus at approximately 2 minutes. (Modified from *Radiology* 147:797–803, 1983).

occurs only with the rapid injection of contrast media. This is a transient phase seen for only 40–60 seconds. The second or nonequilibrium phase occurs approximately 1 minute after the bolus administration of a relatively large volume of contrast material. The final or equilibrium phase occurs approximately 2 minutes after bolus injection and persists until the end of the infusion. At this point, there is a negligible arteriovenous iodine difference. A bolus technique produces maximal opacification of both vessels and parenchyma. It is during the bolus effect and nonequilibrium phase that the most clinically useful information is obtained.

Dynamic scanning of the abdomen using a General Electric 9800 CT scanner is defined as a 2-second scan time with a short (approximately 6–8 seconds) interscan delay. **Nondynamic scanning** is performed with a 2-second scan time and a longer interscan delay. It is probable that with continued improvement in scanners and power injectors, protocols will continue to change. However, basic routine scan protocols will follow.

A. Hepatic CT. Noncontrast examination of the liver is the easiest and simplest way to evaluate the organ. Unfortunately, in a busy practice, a set of images prior to IV contrast administration may delay expeditious evaluation of many patients. Furthermore, most radiologists are uncomfortable interpreting these unenhanced images, especially when differentiating small, peripheral vessels from lesions. This differentiation requires an examination after IV contrast administration. We presently reserve precontrast examinations for evaluation of a known liver lesion in a patient allergic to intravenous contrast and for hypervascular metastasis (e.g., carcinoid, islet cell tumors). Fortunately, the hepatic metastases from the common primary sites (colon, lung, breast, pancreas) are usually hypovascular relative to hepatic parenchyma.

We presently advocate dynamic sequential bolus CT for the evaluation of the liver. Two-second scans are obtained with an 8-second interscan delay, with 10-mm collimation and 10-mm table incrementation. Contrast is injected via the automated injector if at all possible.

Single-level dynamic IV contrast examinations are reserved almost exclusively for suspected hemangiomas. In this instance, multiple timed images at a single level are performed while IV contrast is rapidly administered. After a precontrast evaluation of the liver, the optimal level is chosen. At this level, during contrast injection, approximately six 2-second scans with an 8-second interscan delay can be obtained in 1 minute. The remaining liver can be rapidly evaluated by using incremental dynamic technique. Generally, at the chosen level, delayed 2-, 3-, and 5-minute scans also are obtained. Delay is essential to assess complete enhancement of the hemangioma. Often, an MR examination is obtained on patients with suspected hemangiomas rather than perform the CT evaluation.

When CT portography is performed, contrast is administered via the superior mesenteric artery. A total of 100–150 mL of contrast are infused arterially at a rate of 1–3 mL/sec. Dynamic incremental CT is then performed 5–10 seconds after initiation of the injection and continues during the injection. Approximately 7–11 images/min are obtained.

B. **Pancreatic CT.** The technique for pancreatic CT is the same in terms of dynamic contrast administration. However, scanning is performed by using thin collimation and narrow table incrementation. Scanning is begun at the level of the crossing duodenum or slightly lower, depending on whether the uncinate process is present below this point. During the rapid administration of IV contrast, scanning is performed dynamically in a cephalad direction.

C. **Biliary tree and gallbladder.** The technique that we employ for evaluating the bile ducts and gallbladder is the same as that we use for evaluating the pancreas. Incremental bolus and dynamic scanning are performed by using thin collimation and small table incrementation.

D. **Routine abdomen.** Our routine abdominal CT is generally performed in patients with lymphoma, as yet undiagnosed disease, or in patients referred to rule out intra-abdominal or pelvic abscess. In these patients, the abdomen is scanned from the xyphoid process to the pubic symphysis with 10-mm collimation and 15-mm table incrementation. Occasionally, contiguous 10-mm images are obtained through of the liver. Contrast is injected at a rate of 1 mL/sec for 20 seconds and 0.8 mL/sec thereafter. This is in distinction to hepatic dynamic CT where contrast is initially injected at 2.5 mL/sec for 20 seconds and 1.2 mL/sec thereafter.

E. **Pelvic CT.** If only the pelvis is being scanned, images are obtained from the iliac crest to the pubic symphysis at contiguous 10-mm intervals. Generally, 100 mL of contrast are sufficient for scanning this limited portion of the body.

Contrast is injected at 1 mL/sec for 20 seconds and 0.8 mL/sec thereafter. The aim in these patients is to differentiate pelvic vessels from enlarged lymph nodes. A recent report suggests a delayed technique of scanning the pelvis may be preferable to rapid scanning to optimize enhancement of the pelvic veins (*Radiology* 175:683–685, 1990). The number of possible imaging protocols is infinite, and the use of recuts may be necessary with any technique to clarify questionably abnormal areas.

F. Renal CT. When studying the kidneys with CT, we use 5-mm intervals and 5-mm collimation. Contrast is injected rapidly for 20 seconds, and a slower rate is used thereafter for a total of approximately 150 mL. Even narrower (3-mm or 1.5-mm) collimation may be obtained to better define a cystic mass or suspected angiomyolipoma. In the evaluation of urothelial lesions this dynamic technique frequently will not demonstrate contrast in the collecting system. In these cases, recuts must be obtained. This would include evaluation of suspected urothelial neoplasms, blood clots, and other causes of collecting system filling defects. Precontrast imaging must be obtained in the evaluation of suspected calculus disease.

G. Chest CT. A routine chest CT performed to evaluate disease in the mediastinum or hilar regions is performed by using approximately 150 mL of contrast. Contrast is injected at 1 mL/sec for 20 seconds, at which point scanning is begun. Contrast is injected at 0.8 mL/sec thereafter. The lungs are imaged by using contiguous 10-mm images from the lower neck to the costophrenic angle, and the adrenal glands are included if the patient has lung carcinoma. Alterations to this protocol may be necessary if there is suspected hilar disease.

In the evaluation of a thoracic aortic dissection, initial pre-IV scans are performed from the arch to the diaphragm. At the level of the left pulmonary artery, six images at the same level are obtained with minimal interscan delay during a 20-second, 50 mL IV bolus. Scanning is begun after 25 mL have been injected. After the bolus has been given, a rapid drip of 100 mL of contrast should be administered. After dynamic scanning at the level of the left pulmonary artery is completed, dynamic scanning of the chest by 1-cm intervals should be performed to at least the diaphragm. The scan should be carried into the abdomen to the bifurcation by 15-mm intervals if the intimal dissection extends below the diaphragm.

X. Management of contrast extravasation

A. Incidence. It is estimated that accidental subcutaneous extravasation occurs in as many as 11% of children and 22% of adults receiving intravenous fluids. Most injuries involve small volumes and are not clinically significant. A minimal amount of swelling and erythema may develop, and this usually subsides without untoward side effect. Importantly, severe tissue necrosis can result from even small-volume (less than 10 mL) extravasations, in which case surgical intervention may be required.

Given the frequency with which intravenous contrast is administered, most radiologists will encounter a patient in whom extravasation of contrast media occurs. Extravasations have been reported to occur in 4.5–78% of patients undergoing lower extremity venography. The second most common complication of peripheral venography is that of skin sloughing, comprising 6% of the venographic complications. In a recent study from our institution, five patients had large-volume extravasations of approximately 14,000 body CT examinations performed. The incidence of skin ulceration or necrosis resulting from extravasation of contrast media is difficult to determine. This is due to suspected underreporting in the radiologic literature.

B. Mechanisms. A variety of possible mechanisms can explain the primary toxicity of extravasated contrast to surrounding tissue. This includes **hyperosmolality, ischemia, direct cytotoxicity, mechanical compression, and superinfection.**

C. Presentation. Once an extravasation occurs, nearly all patients show the same signs and symptoms regardless of the agent involved. An immediate stinging or burning sensation is common. The affected site is usually erythematous, swollen, and tender. The initial examination may **underestimate** the severity of an extravasation injury; occasionally, the most severe manifestations may not be apparent for 2–3 days.

D. Treatment. Most research on the medical treatment of extravasation injury has focused on minimizing tissue damage caused by a wide variety of chemotherapeutic agents. **The potential antidotes including the local application of ice, the local injection of isoproterenol or propranolol, local injection of steroids, and application of dimethyl sulfoxide have shown mixed to poor results.**

There are only a few studies on the medical treatment of

radiographic contrast media extravasation, and the results have not been encouraging. Locally injected hyaluronidase has shown mixed results. Local dilution at the entry site with saline or water has not been of help. Local injections of corticosteroids or beta-blockers do not appear to reduce damage to a clinically significant extent.

The definitive treatment of extravasation injuries consists of surgical drainage. Emergent operative drainage was performed on four patients with large volume extravasations of contrast media (*Journal of Hand Surgery* 13:395–398, 1988). Each patient recovered uneventfully, with the authors concluding that all extravasations exceeding 20 mL should be treated surgically within 6 hours. It is possible, however, that some of these patients may have recovered even in the absence of such aggressive therapy.

It is difficult to recommend a precise course of action after a contrast media extravasation. Elevation of the affected extremity is safe and may reduce swelling. Local application of heat or cold may not affect morbidity, but the former may hasten resolution of pain and swelling. **We do not advocate the local injection of any potential antidote.** Patients should be followed closely by both the radiologist and the referring physician until signs and symptoms resolve, since the severity of an extravasation injury may not be initially obvious. **Surgical intervention should only be considered if pain, swelling, or erythema persist or even worsen over time (24–48 hours).** Since operative intervention may be required, prompt consultation with a plastic surgeon is prudent.

E. **Prevention.** Prevention is the most effective means of limiting the severity of extravasation injury. Injections over the dorsum of the hand, foot, or ankle should be avoided. Multiple injections into the same vein are not advocated. Plastic cannulas are preferred over metallic needles, since inadvertent puncture of the back wall of the vein is more likely with metallic needles. IV lines should be easily inspectable so that close observation can be performed during the administration of contrast. It has also been recommended that IV lines not be placed over large tendons, nerves, or blood vessels (*American Journal of Surgery* 137:497–506, 1979).

Indwelling IV lines should generally not be used for contrast media injections unless absolutely necessary. Aspi-

Table 6.3. Patients at Risk for Contrast Extravasation

Elderly and pediatric patients
Unconscious patients
Radiation therapy to IV site
Arterial or venous insufficiency
Peripheral vascular disease
Diabetes mellitus
Raynaud's phenomenon

ration of flowing blood through an indwelling line does not guarantee an adequate position. Extreme caution should be exercised in patients considered at increased risk for extravasation injury (Table 6.3). It is our recommendation that nonionic contrast media be used in patients considered at increased risk for extravasation injury. Nonionic contrast in an animal model has been shown to be less injurious to subcutaneous tissue. Recent experience with large-volume extravasations with nonioinic contrast media have all resolved spontaneously without significant morbidity (*Radiology* 176:65–670, 1990).

Chapter / 7
Vascular Contrast Media Use in the Central Nervous System

Sven Ekholm

The use of contrast media (CM) was an absolute prerequisite in the pre-computed tomography era to gain any useful information about the central nervous system (CNS). Intrathecal injection of air allowed demonstration of the ventricular system as well as the fluid spaces around the brain and cord. To delineate lesions of the cerebral vasculature, it was necessary to inject CM directly into the carotid arteries, which was first done in 1927 by Moniz. The first injection of CM was unsuccessful from a diagnostic point of view, since the agent, 70% strontium bromide, did not result in a useful opacification of the vessels. The idea was born, however, and by changing the injection technique Moniz was able to demonstrate the vessels, although the patient died. Because of this apparent toxicity he changed to iodine salts and soon published the first angiographic image of an intracranial lesion, a pineal tumor in a young patient. The technique developed very fast during the 1930s and 1940s, and it was during this period that most of the diagnostic criteria still in use were described. CM improved somewhat during these years but were still very toxic in the concentrations needed. It was not until the early 1950s, with the development of the first triiodinated CM, when a major improvement in opacification and tolerability was achieved. At the same time, another important milestone was reached, the introduction of the Seldinger technique. The combination of the two finally made the angiographic technique more accessible to the general radiologist.

I. **Contrast media**
   A. **Ionic contrast media.** The modern ionic CM are all developed around the triiodinated benzoic acid molecule using sodium or meglumine as the cation. Meglumine (methylglucamine)

is the common cation in all neuroangiographic CM of this group. One disadvantage in common for all of them is their hyperosmolality with regard to plasma, and this is also the single most important factor for all adverse reactions experienced with these agents. As a consequence, attempts were made early to change the molecule and reduce osmolality. The first dimer developed, Dimer-X, was somewhat better, but it did not gain any significant spread and was soon followed by another, less hyperosmotic dimeric CM, Hexabrix, which contained only one cation to the six iodinated anions. This molecular structure resulted in a marked reduction in osmolality with regard to iodine concentration, but it is still an ionic contrast medium. As a result of the lower osmolality, the patient experiences less discomfort during the injection, with less pain and heat sensations.

B. **Nonionic contrast media.** The real advance in CM was the development of the first nonionic, water-soluble contrast medium, Amipaque. In the Amipaque molecule the side chain containing the cation has been exchanged for a glucosamine. This resulted in a low osmolar monomeric contrast medium intended for angiography and myelography. It has been replaced by newer, less expensive, and easier-to-use agents (see Chapter 10). Today, there are several new nonionic monomers and, lately, nonionic dimers (see Chapter 1) available.

II. **Pathophysiology of contrast media.** CM for neuroangiography have one purpose, to enhance the vascular bed and visualize lesions fed by these vessels. This is the only purpose, and they should ideally be inert in every other aspect. Currently used CM have a very low toxicity, and the incidence of adverse reactions is small in relation to their extensive use. When used for neuroangiography they are accompanied by two categories of reactions: one is related to **idiosyncratic reactions** ("allergy"), which are covered in Chapters 2 and 14; the other includes **physicochemotoxic reactions** where most problems have been related to hypertonicity, which will first have an impact in the vicinity of the injection. The local effects depend on the volume used and the rate of injection. When the CM are injected into a vessel, this immediately results in a very high CM concentration in the blood, affecting first the plasma and blood cells. Second, there will be damage to the vascular endothelium and the surrounding interstitial tissue.

A. **Blood.** Blood can be looked upon as a suspension of cells in plasma that vary in viscosity at different shear rates. Its apparent viscosity is a function of cell deformation, aggregation, and concentration as well as plasma viscosity at a given temperature. The plasma viscosity is primarily a function of protein concentration, and there is no apparent change within normal variation in shear rates. Change in viscosity of whole blood is thus primarily related to the concentration of red cells and their internal viscosity. Increase in red cell concentration will increase hematocrit, reduce the suspending fluid volume, and increase apparent viscosity. The red cells are normally very deformable and can change their shape to a paraboloid with hollow base. This transformation is a normal phenomenon resulting from increased shear rates but also allows the cell to pass through vessels with a diameter as small as 2.9 µm without hemolysis. The deformability will allow blood with a hematocrit of more than 90% to flow when rigid, undeformable red cells with a hematocrit approaching 60% would make blood behave as a solid. The deformation of red cells caused by increasing shear rates will break up the normal formation of aggregates and rouleaux, reduce the volume of immobilized external fluid, and minimize the disturbance of external stream lines. As a result, there is a decrease in apparent viscosity, which also decreases by means of lowered hematocrit when the vessels reduce their diameters from 300 to about 8 µm. Obviously, any factor that influences red cell deformability will change the physiologic behavior of the blood (*The Red Blood Cell*. New York: Academic Press, 1975, Vol. 11, pp. 1031–1133). When strongly hypertonic solutions like ionic CM for angiography are injected into a blood vessel, an osmotic extraction of fluid from cells of the blood, the endothelium, and the surrounding interstitial tissue will result. The osmotic gradient will make the red cell shrink and turn rigid. The mean corpuscular hemoglobin concentration will immediately increase, rapidly increasing the internal viscosity of the red cell. The high viscosity of the CM will also raise plasma viscosity somewhat, but this is counteracted by a sharp rise in plasma volume related to the hyperosmolality of ionic CM. **To summarize, one can expect a more pronounced effect on red cell viscosity from the hyperosmolar ionic CM than from the new, low osmolar nonionic agents.** At the same time, there is more plasma volume expansion from the ionic CM, and they are less viscous than the nonionic CM, but the effects on the red cells are far more important for the apparent viscosity of blood than the effects on plasma.

The intact, deformable red cells will normally pass the capillary bed faster than will plasma, since they are restricted to a smaller volume within the vessel. The effect of hypertonic CM on the red cell is a shrunken, rigid cell body that cannot reshape and pass the capillary bed, resulting in a precapillary packing of such red cells (*Circulation Research* 31:590, 1972; *Acta Radiologica* 362 (Suppl.):123, 1980). This deformation is most often reversible when modern angiographic technique is used, and shortly after the CM injection, the cells will have regained some of their flexibility so they can pass through. As stated above, most of this effect on red cells is related to osmotic effects, but some deformation can be the result of intrinsic chemotoxicity as seen with Amipaque. This deformation (echinocytes) is always reversible in contrast to the hyperosmotic changes (desiccocytes). The origin of this intrinsic chemotoxicity is not entirely understood, but it is less of a problem with the newer, nonionic CM, such as Omnipaque.

The deformed red cell will thus create circulatory problems secondary to their rigidity and inability to pass through the capillary bed. They will, however, not cause any clot formation, since the deformed cell membrane has a reduced and uneven surface area that minimizes the chances of bridging and aggregation that normally take place (*Acta Radiologica: Diagnosis* 19:766, 1978). The increased aggregation that has been described from the use of ionic CM was most likely precapillary packing of red cells rather than aggregation. Furthermore, following the injection of ionic CM there is a sudden decrease in coagulation, with a maximum decrease occurring after about 5 minutes, but it will return to normal within 4 hours (*AJR* 102:441, 1968). There has also been shown a transient decrease in platelet aggregation, and in animal studies these CM have also been shown to enhance the effect of heparin with a subsequent prolongation of clotting time (*Acta Radiologica: Diagnosis* 23:401, 1982). Also, Omnipaque and Hexabrix will inhibit blood clot formation.

From clinical use it has been known that blood that has been inadvertently aspirated into a syringe containing ionic CM did not result in a formation of aggregates. This is in accordance with what is said above about the effects on red cells from the ionic CM. The new, low osmolar, nonionic CM have been said to be more risky to handle in this aspect, since aggregates of red cells have been shown to develop when blood is mixed with these CM in a syringe (*Radiology* 162:621, 1987). However, these aggregates are most likely rather loose and should disperse during injection as they do by shaking the syringe. This apparent drawback of the new, less toxic CM should rather be seen as a proof for their relative inertness, since they allow the normal formation of aggregates. **One should still avoid aspiration and mixing of blood and CM, and because of this, it is recommendable to use a transparent tubing between the catheter and the auto injector syringe.**

B. **Vascular bed.** Besides the increase in plasma volume and subsequent distension of vessels from hyperosmotic CM described above, there are direct effects on the vessel wall causing vasodilatation but also some endothelial damage locally. Chemotoxicity may have some part in the vasodilatation, since this can be created by the low osmolar, nonionic CM but not by equiosmolar saline (*Acta Radiologica* 362 (Suppl. A):43, 1980). Whatever the origin of vasodilatation, it seems to be a direct effect on the vessel wall, and it is identical in most vascular regions with the exception of the kidney. Thus, injection of hypertonic CM in the common carotid artery will result in dilatation of the intracranial as well as the extracranial vessels, but it may also result in a shunting of blood from arterioles to venules intracranially (*Contrast Media in Radiology*. New York: Springer Verlag, 1982, pp. 115–122). If there now is a severe stenotic lesion affecting only the internal carotid artery, the injection of hypertonic CM may possibly create additional circulatory disturbances intracranially as a result of increased blood flow through the dilated extracranial vessels. This is secondary to the higher concentration of CM in the extracranial vascular bed under these circumstances with a subsequent reduction of intracranial blood flow. The new, low osmolar CM should be less deleterious in this aspect. Moreover, since they cause less pain and heat sensation, they will reduce the risk for motion artefacts, which is most important when the digital subtraction angiography (DSA) technique is used.

There is at least one exception to the rule of CM vasodilatation of intracranial vessels. In patients with acute subarachnoid hemorrhage, vasospasm is a common finding in areas with large collections of fresh blood in the subarachnoid space. Those vessels affected by vasospasm are assumed to be more sensitive to ionic CM that may prolong and/or enhance the spasm and thus further reduce the blood supply to suffering regions. In animals it has been shown that spasm induced by a mechanical trauma will be significantly prolonged if ionic CM are injected immediately following the initiation of spasm (*Journal of Neurosurgery* 17:1055, 1960). If spasm is noted during the angiographic search for aneurysm, it is probably advisable to postpone the examination if ionic CM are used. The nonionic CM seem to create less problems with spasm. Still, if the examination is continued, the number of injections should be reduced to a minimum to detect the aneurysm. A complete examination of all intracranial vessels to rule out further aneurysms can be accomplished at a later, more suitable opportunity when spasm is no longer present.

The blood vessels of the CNS, in contrast to the vessels of most other organs, have an endothelium with a continuous basement membrane with cells connected by tight junctions. The resultant barrier between the capillaries and the CNS, called the blood-brain barrier (BBB), will hinder the passage of many solutes in both directions (see Chapter 10). This protection of CNS by means of a BBB is fully developed at birth, and there is no available evidence that it will be less effective with age. There are a few areas that normally lack this barrier. The reason for this is not entirely known, but it may be a way for the brain to "sense what is going on outside" for immediate adjustment and protection of the body. One example of this could be the area postrema of the brainstem, which most likely is responsible for the rather common sensation of nausea and vomiting from ionic agents as a result of CM leakage into this area following intravascular injections (*Annals of Neurology* 13:469, 1983).

Some diseases will cause a damage to the BBB, e.g, meningitis, and in others the barrier is incomplete, e.g., gliomas, or entirely missing, e.g., metastasis. The lack of BBB in metastasis is related to their origin, and the vessels of metastasis are identical with the primary tumor. In all those lesions, more or less CM will leak out of the vessels and concentrate locally in the tissue and in this way enable their detection. CM, which in this way have passed the protective BBB, will come in direct contact with the neural tissue and can reach a concentration that may interfere with the normal neuronal function. There are several studies regarding this potential problem. **It has been shown that patients with metastatic brain lesions were stricken with seizure in a frequency that varied between 6% and 19% of examinations with ionic CM (*AJR* 140:787, 1983).** This should be compared with about 0.01% in the general population. Another area of debate concerns the acute brain infarction where CM will leak out and may possibly cause a permanent damage to the tissue in the penumbra zone and thus enlarge the area of infarction (*Neuroradiology* 19:241, 1980). Although there may be doubts about this damaging effect of some CM, one should always try to use the least neurotoxic agent.

Ionic CM or other compounds with similar osmotic effects will cause a damage to the BBB if injected over a certain time period. Most of the damage of the CM is an osmotic effect, but there is also a contributary chemotoxicity, since CM isoosmolar to saline or mannitol solutions will result in more damage to the BBB (*Annals of Neurology* 13:469, 1983). Accordingly, this detrimental effect of hyperosmolality is less of a problem with the new nonionic CM. The time span between consecutive injections of nonionic CM in carotid angiography can thus be shortened without risking an additive damage to the BBB, which has been described with ionic CM if injected too fast. Questions have also been raised regarding the impact of prolonged high serum concentrations of CM. One study in sheep made anuric immediately following the injection of ionic CM (4 mL/kg body weight) and followed over 2 days did not support this theory in otherwise healthy animals where serum osmolality reached a maximum of 380 mmol/kg (*Acta Radiologica* 29:237, 1988).

C. **Contrast media neurotoxicity.** The neurotoxic properties of the ionic CM are well established, but in spite of this, intra-arterially administered ionic CM rarely result in neurotoxic effects. As mentioned above, there are certain disease processes that may increase the seizure incidence, and seizures have been reported in 0.2% of the patients following cerebral angiography when ionic CM were used. Most of those patients had a deficient BBB secondary to brain tumors. Another complication is complete or incomplete **cortical blindness** following **vertebral angiography.** This has been described in 0.3–1.0% of the cases, but most of these problems are transient and disappear in a period of hours to a few days. The origin is thought of as a direct cortical toxic effect, and it has been speculated that the BBB of the visual cortex is more fragile and liable to damage than other cerebral areas (*Journal of Neurosurgery* 54:240, 1981). Another region of concern is the spinal cord where aortic injections as well as selective spinal angiography with ionic CM have been reported to result in spinal cord injury with permanent deficits in some patients. This again is a much smaller problem with the new, nonionic CM where most problems have been related to catheter technique. In summary, it seems obvious that the new nonionic CM, although not inert, are less prone to cause neurotoxic effects (see Chapter 10).

III. Neuroangiography

A. **Indications for neuroangiography.** The introduction of CT altered many of the indications for neuroangiography. In many cases, such as head trauma, angiography is no longer necessary, since CT will provide all information needed to evaluate, for example, a subdural hematoma and will add more information about concomitant parenchymal damage. The only indications left for angiography in head trauma are suspicions of an arterial laceration. Also, a blunt trauma of the neck vessels may lead to the formation of a pseudoaneurysm or an intimal dissection that can result in a vascular occlusion. Other important indications left are: (1) **subarachnoid** or **intracerebral hemorrhages** of nontraumatic or nonhypertensive origin to visualize possible aneurysms or other vascular malformations; (2) **transitory ischemic attacks and reversible ischemic neurologic deficits,** to evaluate if there is a stenotic and/or ulcerative lesion of the carotid arteries that may lend itself to surgical correction; (3) **subclavian steal phenomenon;** and (4) **arteritis.** In tumor cases, angiography has lost most of its past importance and is now mainly used preoperatively to get a better understanding of the relationship between tumor and adjacent vessels. Such lesions include, for example, pituitary tumors, tumors in the vicinity of the middle cerebral artery bifurcation, and falx meningiomas along the superior sagittal sinus. Today, angiography is also performed to embolize tumor feeding arteries preoperatively and for other interventional procedures. In the future, one can expect interventional angiography to become even more important and, eventually, the major indication for neuroangiography.

B. **Angiographic preparation of the patient** is, for the most part, common sense. The patient history and laboratory data as well as pertinent earlier radiographs should be carefully evaluated. The physician performing the procedure should always try to meet the patient before the examination and give a review of why it is done, what potential hazards are involved, and what physical sensations the patient can expect to experience during the procedure. If this is done in a proper manner, the relationship between patient and physician will improve, thereby reducing the anxiety experienced by most patients. The patient should be well hydrated and asked to void just before the examination is begun. Sedation is usually not needed, but diazepam can be used in extremely anxious and tense patients.

C. **Contrast media injection technique.** The complete angiographic technique is not covered in this text, but a short summary of CM injection recommendations follows. In the past, when conventional film technique was the only available option, the total CM dose was larger than what is needed with intra-arterial DSA. DSA was initially advocated as an intravenous technique to shorten and make the angiographic examinations easier but also less risky than conventional arterial angiography. Initially, IV DSA was commonly used to screen patients with suspected atherosclerosis of the common carotid artery bifurcation. Some hazards with the catheterization can be avoided with this technique, but because of the dilution there is a need of a relatively large CM volume. This has limited the use of modern, low osmolar CM, since they are quite expensive. On the other hand, high-speed injections of large volumes of ionic CM in the vicinity of the right atrium have resulted in a not insignificant number of complications. Rapid injections like that in this region will cause a fall in the systemic arterial pressure and an increased pulmonary arterial pressure. Furthermore, the lung tissue is rich in vasoactive substances that are more readily activated from the injection of hyperosmolar CM. This potential release of vasoactive substances may also explain the higher incidence of "allergic" reactions following intravenous injections compared with intra-arterial injections. Moreover, since intravenous DSA is most often applied in older, more fragile patients, these effects may be even more serious (*Journal of the Canadian Association of Radiology* 31:8, 1981). The general quality of intravenous DSA is also rather poor, most often because of discomfort during the injection resulting in patient motion.

The use of intra-arterial DSA has, on the other hand, become more and more popular with the continued improvement in image resolution. Very soon, the quality of DSA equipment will have reached a level where conventional film technique can be dropped. The instant review possibility with DSA will shorten the total catheterization time, and the superior CM detectability will reduce the CM load to the patient. A suggestion of CM injection technique can be found in Table 7.1.

D. **Contraindications to neuroangiography.** There are no absolute contraindications for angiography when the examination is vital to the patient. There is an increased morbidity in the extremely ill and comatose patients, but this morbidity seems to be lower with the new, low osmolar CM. There are, however, some patients where extra caution is recommended. First, patients with known "allergic" reactions to iodinated CM should be premedicated with steroids and carefully watched during the procedure. The steroid regimen should start 2 days before the examination with 30 mg prednisone. This is repeated the following evening and once more on the day of examination. Caution is also recommended in patients with migraine headache. If the patient is anuric, hemodialysis should start promptly following the angiography, but a filter suitable for CM should be

Table 7.1. Suggested CM Injection Protocol

Vessel[a]	CM Concentration (mg I per mL)		Volume (mL)[c]	Injector Rate (mL/sec)[c]
	Conventional Film	i.a. DSA[b]		
Aortic arch	280–350	140–180	50 (30)	30 (20)
CCA	280–350	140–180	10 (8)	10 (8)
ICA	280–350	140–180	7 (6)	7 (6)
ECA	280–350	140–180	10 (6)	3 (2)
Vertebral				
Selective	280–350	140–180	6 (4)	5 (3)
Subclavian[d]	280–350	140–180	30 (20)	10 (7)

[a]CCA, common carotid artery; ICA, internal carotid artery; and ECA, external carotid artery.
[b]i.a. DSA, intra-arterial digital subtraction angiography.
[c]The number in parentheses relates to DSA technique, assuming high-quality equipment.
[d]Catheter in the midportion of the subclavian artery and a blood pressure cuff to block peripheral flow to the arm during the injection.

used. In marked hypotension or when there is severe spasm of the intracranial vessels, the cerebral blood flow is retarded, resulting in stasis of CM. This will increase contact time and thus the risk for damage to the BBB with subsequent leakage of CM into the CNS, especially when hyperosmolar, ionic CM are used. The new, low osmolar CM are definitely less harmful to the BBB as well as the CNS, but caution should still prevail.

E. Choice of contrast medium. If selection of CM is based on economy alone, the only choice is the old ionic agents. There are, however, many advantages with the new, low osmolar CM, which in my opinion are far more important. The following comparison is basically a summary of what has already been said in the text: (1) The effects on the red blood cells (RBCs) are much milder with low osmolar CM, which will result in less precapillary packing. (2) The low osmolar CM cause less vasodilatation and flow disturbances while the risk of intracranial pressure drop should be smaller in patients with severe stenosis of the internal carotid artery. (3) The low osmolar CM cause less endothelial damage, reducing the risk of BBB damage. (4) The nonionic CM are less neurotoxic. (5) The nonionic CM have a lower incidence of "allergic" (idiosyncratic) reactions. (6) The low osmolar CM create less pain and heat sensation, which will reduce patient motion and improve image quality, especially with DSA technique. (7) Nausea and vomiting are less of a problem with the low osmolar CM.

One of the few potential problems with the low osmolar CM when compared with the hypertonic CM is their tendency to cause aggregates of RBCs when blood is aspirated and left for a while in a CM-containing syringe. The real hazard of this is still debated, but it can be easily avoided by using a large, transparent tubing between the catheter and injector syringe. This tubing will reduce the risk of inadvertent aspiration of blood into the syringe.

Chapter / 8
General Angiography

Michael A. Bettmann

I. **General considerations.** Angiography involves percutaneous placement of a catheter with subsequent injection of contrast and filming for visualization of major arterial or venous structures. The angiography dealt with in this chapter refers to the abdominal aorta and its branches. This, in turn, involves direct injection of contrast media via a catheter, utilizing rapid manual injection with a syringe or mechanical injection with a programmable mechanical injection apparatus known as a power injector. While the types of contrast agents utilized are the same as those used for all other intravascular applications, the method of delivery of the contrast and the speed of delivery are unique.

II. **Visceral angiography.** Visceral angiography involves the visualization of the abdominal aorta and its branches, including the celiac axis, the superior mesenteric artery, the inferior mesenteric artery, and branches of these vessels. Visualization of the renal arteries is generally referred to separately but falls into this category in terms of technique. For visualization of the abdominal aorta and its branches, either anteroposterior or biplane (with frontal and lateral views) filming is generally utilized. A pigtail catheter is usually used, and contrast is injected with the pressure injector at a rate ranging from 10 to 30 mL/sec, with a 1–3-second injection duration and a total volume of 25–50 mL. Filming is generally performed at an initial rate of 2–3 films/sec. For abdominal aortography, a contrast agent with 300–376 mg of iodine per mL is employed.

For selective visceral angiography or for renal angiography, preformed curved catheters are generally utilized. Contrast is injected at various rates, depending on the vessel injected. Volume and rate depend on the indication, patient anatomy, cardiac output, and flow through the vessel. For example, in a study performed to identify a bleeding source in a patient in shock, the heart rate will be rapid but cardiac output may be decreased. There may be a marked decrease in visceral flow due to vasoconstriction. In a patient with a large vascular tumor, blood flow may be extremely rapid, and a high rate and volume of injection may be necessary. For celiac axis visualization, 6–10 mL/sec for 6–10 seconds is the usual range. For injection of a main renal artery, an injection rate of 5–8 mL/sec for a total of 10–15 mL is usual. This is generally done by pressure injector, as with celiac axis injection, because of the relatively high flow rate. For injection of the superior mesenteric artery, rates and volumes similar to those utilized for the celiac axis injection are employed. For the inferior mesenteric artery, injection is either by hand or with a pressure injector, at a rate of 2–4 mL/sec, with a total injection volume of 4–6 mL. For all visceral angiography, filming is most often done with 14 × 14-inch cut film and a rapid film changer. The iodine content varies from 282 to 370 mg/mL; the lower concentration is generally more comfortable for the patient and thus leads to less patient movement. In all but the largest patients or those with extremely rapid flow, an iodine content of 282 mg/mL provides diagnostic vascular opacification.

III. **Peripheral angiography.** Peripheral angiography is most often performed for evaluation of the extent, distribution, and severity of atherosclerotic disease. It is also often useful in the evaluation of trauma. It is rarely utilized for other indications. To visualize the entire arterial system from the aortic bifurcation to the feet, a large volume of contrast is injected at a relatively slow rate, with filming accomplished either with a long leg film changer or a moving tabletop. Additional areas are subsequently elucidated by utilizing oblique views and/or stationary positioning. For the major angiographic runoff studies, injections are generally made at a rate of 5–10 mL/sec for a total of 60–100 mL. With a stationary long leg film changer, volumes can be slightly less. Similarly, if only one leg is studied, volumes are just over half, and if a stationary view is utilized, volumes are substantially decreased, although rates remain in the range of 3–8 mL/sec. Most investigators prefer to use contrast agents at a concentration of 370 mg of iodine per mL, although a concentration of 280–320 mg of iodine per mL is also often used. Visualization of vessels with this lower concentration is almost invariably diagnostically adequate but not as clear as with the higher concentration. **A key consideration in peripheral angiography** is that the injections are often painful, unless digital subtraction imaging is utilized. This pain is discussed further in **Section V, Complications.**

IV. **Digital subtraction angiography**

A. **Indications** Digital subtraction angiography (DSA) has evolved from initial widespread use with intravenous injections and suboptimal visualization in a high percentage of cases, to intra-arterial use. As compared to cut-film angiography, DSA provides an inherent improvement in contrast resolution but a loss of spatial resolution. For most purposes, this loss of spatial resolution is more than offset by the gains. With DSA the basic principle is that a mask image is subtracted from a contrast-containing image, to allow visualization of the contrast-filled structures only. To accomplish this, there must be minimal movement between the time the mask and the subtracted images are obtained. That is, patient cooperation is imperative. This limits the utility of DSA in certain patients, such as those with uncontrollable tremors, an inability to comprehend or cooperate, or other reasons for an inability to remain perfectly motionless for brief periods of time.

The two major advantages of DSA are the improved visualization of structures that fill poorly with contrast, such as distal vessels that fill only via collaterals, and the ability to use dilute (i.e., lower iodine concentration) contrast material. This leads to a marked decrease in discomfort for the patients during painful examinations, such as peripheral angiography, as will as a lower total volume of contrast and consequently a lower osmotic load.

B. **Techniques and applications.** There are various ways of producing DSA images. These have been discussed previously by Levin (*AJR* 143:447–454, 1984). Most digital systems for peripheral and visceral angiography are limited to 3 or 4 images/sec. Cardiac systems capable of imaging 30 or more frames/sec for a few seconds are available but are more often utilized for direct digital imaging then for digital subtraction imaging, and consequently, full-strength contrast material must be used. Digital subtraction imaging is particularly useful for evaluation of areas in which there is inherently little motion, such as peripheral arteries. Similarly, it is extremely useful in the carotid circulation. It is less valuable in the abdomen for evaluation of visceral arteries because of inherent bowel peristaltic movement and respiratory motion. Some investigators, therefore, routinely use **glucagon** in abdominal DSA. Because of its improved contrast sensitivity, DSA can sometimes be valuable in defining bleeding sites in the abdomen. It is most often used with filming at a rate of 0.5–2 exposures/sec. For visualization of collateral supply in the lower extremity, filming at a rate of 1 frame/sec is employed, with a delay of several seconds in filming following the injection of the contrast, as dictated by the speed of blood flow.

C. **Contrast agents.** Standard contrast agents are utilized for DSA. Because of the improved contrast sensitivity, contrast can be utilized at either a much lower iodine concentration or a much lower volume. In general, investigators utilize dilute contrast, with a concentration ranging from 80 to 200 mg of iodine per mL (e.g., Hypaque 60 diluted to Hypaque 20, or Conray 43). Contrast in this case is injected at the usual rate for the vessel or area to be imaged. Alternatively, a much lower volume of contrast can be injected. For example, for cut-film imaging of the pelvis, a usual injection is 10 or 15 mL/sec for 2 seconds. The contrast is used at an iodine concentration of 280–370 mg/mL. With DSA a total of 8–10 mL of the same concentration is injected over 1 second, or a total of 10–15 mL/sec for 2 seconds is given using dilute contrast material. Again, spatial resolution is superior with cut films, but contrast resolution is superior with either DSA method, and diagnostic information, if there is no patient movement, ranges from adequate to excellent. **The major advantages of the decreased contrast volume or concentration are the decreased osmotic load, decreased patient discomfort and, therefore, patient movement, and lessened need for use of the more expensive low-osmolality contrast media.**

V. **Complications**
A. **General considerations.** The complications encountered in angiography are to a large extent identical with those that occur with use of any contrast injection. There are, however, several specific differences. First, as noted above, patient discomfort is a major consideration with peripheral arterial injections and external carotid injections. It is rarely an important consideration with other angiographic procedures. Second, thrombotic and hematologic effects of contrast are more important with intra-arterial injections. This is because with the use of catheters, there is prolonged contact

between the mildly to moderately thrombogenic surface of the catheter, the blood, and the contrast material. Third, the overall incidence of reactions to contrast media is thought to be less with intra-arterial use then with intravenous use (*Radiology* 143:1–17, 1982). It is not yet clear whether this is true or not. While there is a lower incidence of nausea and vomiting with intra-arterial injections than with intravenous injections, the overall incidence of complications with angiography is greater then with procedures such as intravenous urography or contrast-enhanced computed tomography (G. Ansell and R. Wilkins, eds. *Complications in Diagnostic Imaging,* 2nd ed. London: Blackwood, 1987). Many of these complications, however, are related to catheter use, underlying patient disease, or a combination of the two rather than specifically to the contrast media.

B. **Characterization of complications.** The most common complication of intra-arterial contrast use is discomfort. This is related primarily to osmolality, but as shown by Yahiku and Smith (*AJR* 9:137–139, 1988), there are some specific formulation-related factors that affect pain. As a general rule, pain on injection is marked with conventional contrast agents at an osmolality of 1000 mOsm/kg or greater, is generally mild to moderate with low-osmolality contrast agents at osmolalities of 600–800 mOsm/kg, and is rarely noticed with dilute contrast agents at osmolalities between 300 and 500 mOsm/kg. The relationship between iodine concentration and osmolality for high- and low-osmolality contrast agents is shown in Table 8.1. Thrombotic complications are, in general, related to underlying disease but may be influenced by the contrast utilized (*Catheterization and Cardiovascular Diagnosis* 14:159–164, 1988). It is clear that ionic contrast agents, of both high (e.g., Conray) and low osmolality (e.g., Hexabrix), markedly retard coagulation in a static situation such as a syringe. The effect of nonionic contrast media is less marked. It is not clear at this point whether nonionic agents actually have a peculiar procoagulant effect (*Radiology* 174:459–461, 1990). **It is likely, however, that careful angiographic technique with frequent catheter flushing utilizing heparinized saline is the most important consideration in preventing thrombotic complications.**

Contrast-related renal failure is poorly defined in angiography, as it is with intravenous injections. This is dis-

Table 8.1. Relationship between Iodine Concentration and Osmolality

Contrast Agent	Iodine Concentration (mg/mL)	Approximate Osmolality (mOsm/kg)
Hypaque 76	370	1800
Renografin 76	370	1800
Conray 370	370	1900
Hypaque 60	282	1400
Conray 60	282	1400
Conray 43	212	1000
Isovue 370	370	750
Omnipaque 350	350	780
Hexabrix	320	580
Optiray 320	320	700
Omnipaque 300	300	600
Optiray 240	240	500
Plasma		300

cussed in detail in Chapter 3. Recent evidence suggests that the total volume of contrast is not a particular risk factor for contrast-related renal failure with angiography (*Radiology* 167:607–611, 1988), but preexistent azotemia, particularly in association with diabetes mellitus, appears to be (*New England Journal of Medicine* 320:143–149, 1989; *AJR* 157:66–68, 1991).

Life-threatening reactions to contrast administration may be related to a poorly understood immune mediate response. This is thought to occur, but with intra-arterial as with intravenous injections, more likely and more frequent explanations are **direct cardiac effects, or vagal stimulation with resultant bradycardia and hypotension** (*New England Journal of Medicine* 317:891–893, 1987). As with severe reactions with any contrast administration, it is important to attempt to define the etiology of a reaction (see Chapter 14). If a patient is **hypotensive and tachycardic,** standard resuscitation with oxygen, fluid, and vasopressors, even including epinephrine, is indicated. If the etiology of a life-threatening reactions is an underlying **arrhythmia,** such as ventricular tachycardia, this should be treated directly. If a patient is **hypotensive but has no elevation in heart rate or is bradycardic,** this is likely to be a vagal reaction. This should be treated initially with intravenous fluids and leg elevation. If more severe, it should be treated with intravenous atropine, at a dose of 0.4–1.0 mg. It must be kept in mind that true allergic reactions are extremely rare and their usual manifestation is dermatologic rather than cardiovascular.

C. **Risk factors and prevention of reactions.** Patients who fall into certain categories are particularly likely to experience adverse reactions, generally hemodynamic ones, to contrast media. This includes patients with an **unstable cardiac status,** such as those in cardiogenic shock (*American Journal of Cardiology* 61:1334–1337, 1988), those with **severe congestive heart failure,** those with **marked pulmonary hypertension,** and those with **severe cardiac disease and borderline compensation.** This latter category includes patients with severe although mildly or minimally symptomatic coronary artery disease, as well as those with tight aortic stenosis. Patients with significant **cerebrovascular disease** may also be at particular risk. This includes patients with recent marked head trauma, tumor, or stroke. The reason is that contrast agents of all sorts may be neurotoxic when in contact with the brain substance. Normally, at clinical doses, contrast agents do not cross the blood-brain barrier. When the blood-brain barrier is disrupted, however, as occurs in severe head trauma, contrast agents may come in contact with the brain substance and may cause various reactions. Neural effects on the heart, which are poorly understood, are probably more frequent and more relevant (*American Journal of Cardiology* 60:15J–19J, 1987).

Although a prior significant contrast reaction and a strong history of allergies or active asthma are generally accepted as risk factors for contrast reactions (*New England Journal of Medicine*: 845–849, 1987; *Radiology* 175:621–628, 1990), this relationship is present but less clear with angiography. The debate concerning the use of high- versus low-osmolality agents exists in angiography as with all contrast use (*Radiology* 175:616–618, 1990). It has been established, with intra-arterial administration as with intravenous administration, that the incidence of *mild* reactions is less with low-osmolality as than with high-osmolality contrast media. The difference in incidence of moderate, severe, and fatal reactions, however, is not clear. Hexabrix, the only ionic low-osmolality contrast agent, is rarely used intravenously, as it appears to be associated with an increased incidence of mild reactions such as nausea, vomiting, and urticaria. It has particular advantages intra-arterially, however, both be-

Table 8.2. Indications for Use of Low-Osmolality Contrast Agents in Angiography

---

Acutely unstable cardiac status (cardiogenic shock, evolving myocardial infarction, unstable angina, acute congestive heart failure)

Other severe cardiac disease (e.g., aortic stenosis, severe coronary artery disease, pulmonary hypertension, severe cardiomyopathy)

Significant intracerebral pathology (recent stroke, large tumor, acute head trauma)

Active severe allergies or asthma

Prior significant reaction to contrast agents (e.g., laryngotracheal edema or severe diffuse urticaria, not nausea, vomiting, or a vagal reaction)

Painful examination (e.g., peripheral angiography) (DSA is often an acceptable alternative)

Strong possibility of patient detriment or of loss of information, due to discomfort or minor reaction (e.g., patients with unstable spinal cord fractures, patients unable to cooperate fully)

Marked patient anxiety

---

cause of slightly lower cost and because of a more marked antithrombotic effect, compared with nonionic agents. **The most rational approach at this point is selective use of low-osmolality contrast media in certain patients who are likely to be at particular risk or are likely to gain significant benefits for other reasons (Table 8.2).**

VI. **Future considerations.** DSA will probably play an increasing role in arterial imaging. If appropriately used, this may well lead to lower radiation doses as well as to lower doses of contrast media. The use of dilute contrast agents with DSA obviates the use of the more expensive high-osmolality contrast media in most cases. Percutaneous catheter angiography will remain important, more for therapeutic interventions than for diagnostic purposes. Currently, there is a role for selective use for low-osmolality contrast media in certain patients including those undergoing particularly painful procedures and those in certain high-risk categories. With the development of even safer contrast media on the one hand, and of less expensive low-osmolality agents on the other, there will be continued evolution in contrast use. With the continued improvement in noninvasive modalities for vascular imaging, including magnetic resonance imaging and Doppler ultrasound, more specialized use of angiography with better defined utilization of specific contrast media is likely to emerge.

Chapter / 9
Peripheral Venography

John A. Kaufman
Michael A. Bettmann

I. **General considerations.** Contrast examinations of the peripheral veins are most commonly employed in the evaluation of suspected deep venous thrombosis (DVT). Less frequently, they are utilized to study other venous disorders, such as valvular incompetence. Although this chapter focuses on venography for DVT, the principles of contrast usage discussed here are applicable to contrast studies of the peripheral veins in general.

The role of venography in the diagnosis of DVT is being redefined as sensitive, specific noninvasive techniques of evaluating the peripheral veins are developed (*Radiology* 168:97–112, 1988). When a venogram is indicated in the workup of a patient in whom lower extremity DVT is questioned, it is usually performed according to the method described by Rabinov and Paulin (*Archives of Surgery* 104:134–144, 1972), with or without pressure infusion (*Radiology* 138:730–732, 1981). Upper extremity venograms are performed with hand injection technique. A radiologist or other vascular specialist should be involved in the selection of imaging modalities in the evaluation of patients with venous diseases.

II. **Contrast for venography**
   A. **General considerations.** Venography differs from angiography in that the injected contrast does not replace intravascular blood volume but rather mixes with it. This is due to the large capacity of the venous bed, which can approach 800 mL in a single lower extremity (*Surgery* 8:604, 1940). To opacify such a potentially large volume, phlebographic contrast agents must not only mix well with blood but also have an iodine concentration high enough to produce diagnostic images after dilution. However, as most local complications of venography appear to be related to osmolality, which in

turn is directly related to iodine concentration, dilute contrast agents are desirable (*AJR* 134:1169–1172, 1980). Prior to the development of low osmolar contrast agents, an iodine content of 200–225 mg/mL was demonstrated to provide excellent visualization of venous structures while minimizing local side effects. Generally, volumes of 80–150 mL of contrast are required for a complete study (*Radiology* 122:101–104, 1977). This concentration and volume of contrast work equally well for the newer, low osmolar agents (*Radiology* 165:113–116, 1987). Phlebography performed with digital fluoroscopic systems employs similar contrast materials but somewhat lower volumes (<100mL) than standard equipment (*AJR* 153:413–417, 1989).
B. **Contrast types**
   1. **High osmolar.** Water-soluble contrasts at an iodine concentration of 200–225 mg/mL are well tolerated and opacify veins well. Examples of readily available agents include Conray 43, and Renografin 60 diluted 3:1 with sterile, injectable 0.9% NaCl (normal saline). In very large patients or in the presence of extremely high venous flow, undiluted 60% contrast (282 mg I per mL) can be used but with a higher risk of local morbidity.
   2. **Low osmolar.** Published clinical data are available for a number of agents, such as iopamidol (200 mg I per mL), iohexol (240 mg/mL), meglumine ioxaglate (320 mg I per mL), and metrizamide (280 mg I per mL) (*Radiology* 165:113–116, 1987; 177:503–505, 1990; 147:399–400, 1983; 140:651–654, 1981). The latter two were studied at iodine concentrations higher than current practice, although with fewer complications than the high osmolar agents with similar iodine contents to which they were compared. Low osmolar agents with low iodine concentrations (200–240 mg I per mL) readily produce diagnostic images in the majority of patients.
III. **Contrast-related complications.** The overall incidence of complications reported with venography ranges from 18% to 48%, of which most are minimal, localized, and self-limited. Severe systemic complications, such as cardiorespiratory arrest, occur in less than 1% of patients (*AJR* 151:263–270, 1988). The mortality of undiagnosed and therefore untreated DVT is not known but is most likely linked to pulmonary embolization of clot (PE). The mortality rate from untreated PE is approximately 26% (*Diagnostic Angiography*. Philadelphia: W. B. Saunders, 1986, p. 598), far

higher than the risk of a life-threatening contrast reaction during venography. Specific contrast-related complications are described below.

A. Pain. Pain, variably described as foot or calf discomfort, pressure, or burning, is the most common complication of phlebography. It is limited to the duration of the examination. Using high osmolar contrast with low iodine concentrations (200–225 mg I per mL) reduces the incidence of severe symptoms from 24% to less than or equal to 5%. This is comparable to the incidence of severe pain experienced with the more expensive low osmolar agents (*Radiology* 165:113–116, 1987). Most patients experience only mild pain with either dilute high osmolar or low osmolar contrast. The addition of lidocaine (10–40 mg lidocaine per 50 mL of contrast) has been reported to decrease pain during the examination when high osmolar agents are utilized (*Radiology* 133:788–790, 1979; *Australian and New Zealand Journal of Medicine* 14:622–625, 1984). However, this seems unnecessary with current contrast usage. Conducting the examination in a rapid manner also helps decrease patient discomfort. The etiology of the pain appears to be related to contrast effects on the endothelial cells lining the veins.

B. Extravasation. Local extravasation of contrast is disconcerting to the patient and physician but usually of little or no clinical significance. Careful monitoring of the injection site, attention to patient complaints of localized pain, and use of contrast with low iodine concentrations or low osmolarity all contribute to minimize the incidence and significance of this complication. Small extravasations require no treatment or only minor symptomatic therapy. Large extravasations can be safely managed with local heat, massage, and careful observation (*AJR* 151:263–270, 1988). Patients with poor arterial inflow, or otherwise-jeopardized extremities are at greatest risk for the rare episode of tissue slough (*Radiographic Contrast Agents,* 2nd ed. Gaithersburg, MD: Aspen Publishers, 1989, pp. 202–212). An experienced surgeon should be consulted early in the management of a patient with a large extravasation into already-compromised tissue. There is no indication for the injection of hyaluronidase into an area of extravasation, as this may actually increase the amount of tissue damage (*Radiology* 99:511–516, 1971; 14:622–625, 1984). In summary, no specific therapy has yet proven effective.

C. **Postphlebography syndrome.** This condition is similar in presentation to but distinct from postphlebographic DVT. Approximately 6–12 hours after the venogram, pain and swelling low in the calf have been reported in up to 7.5% of patients studied with dilute contrast agents (*Radiology* 122:101–104, 1977). Occasionally, these symptoms are associated with malaise and fever. Resolution of these findings should be virtually complete by 5 days. Although the etiology of this syndrome is unclear, it is probably related to an inflammatory response to the contrast rather than to thrombosis. However, the threshold for performing noninvasive evaluation, especially ultrasonography, should be low, as this entity is easily confused with DVT or cellulitis.

D. **Postphlebographic DVT.** The development of venous thrombosis after a normal venogram is the most severe common local complication of contrast phlebography. Prior to the generalized adoption of contrast agents with decreased iodine concentrations, the incidence of acute DVT following venography was 26–48%. Two thirds of these thrombi occur in superficial veins, and most are below the knee but may propagate (*Radiology* 140:651–654, 1981). When the concentration of iodine is kept to 200—225 mg I per mL, or low osmolar contrast media are utilized, the incidence of thrombosis drops to 0–9% (*Radiology* 165:113–116, 1987). Infusing 150–200 mL of an IV solution (5% dextrose in water, 5% dextrose in 0.45% saline, or heparinized saline) through the venography system immediately after the examination prevents stasis of contrast in the veins. Prolonged contact between the endothelium and contrast may be important in the initiation of DVT after venography (*AJR* 143:629–632, 1984).

If a patient develops new symptoms suggestive of an acute DVT within a few days of a negative phlebogram, evaluation with a noninvasive test such as duplex or color-flow ultrasound or with I–125 fibrinogen uptake should be performed. The latter examination is sensitive but relatively nonspecific and difficult to interpret if there is a concurrent inflammatory or posttraumatic condition in the extremity. If the results of the noninvasive testing are inconclusive, a venogram should be repeated.

E. **Severe systemic (anaphylactoid) reactions.** Venography patients are at no lesser or greater risk for such reactions than other patients receiving intravascular contrast. This

study should never be performed without the equipment and trained personnel necessary for resuscitation readily available.
   F. **Renal insufficiency.** The accepted risks of contrast administration to patients with normal or impaired renal function apply when venography is performed.
IV. **Indications for specific contrast agents**
   A. **General.** Venography contrast, whether high or low osmolar, should contain an iodine concentration in the range of 200–225 mg I per mL. In selected cases, such as extremely large patients or patients with very high venous flow (i.e., known cellulitis), more concentrated contrast may be utilized with caution (a rapid examination and infusion of intravenous fluids through the venography needle, as discussed under Section III, D. **Complications: Postvenography DVT,** are recommended). In most instances, the selection of a particular contrast agent can be guided by the general principles of contrast usage. Almost all patients, with no specific risk factors, can be studied with a dilute, high osmolar contrast such as Conray 43, without significant morbidity. Patients with a prior significant contrast reaction require adherence to the protocols outlined in Chapters 2 and 14. In all cases, a diagnosis should be sought with noninvasive means first, before contrast venography is considered. Listed below are a few selected guidelines for situations not addressed in this paragraph or elsewhere in this book.
   B. **Unknown/unobtainable history.** This category applies to patients in whom an adequate determination of risk factors is impossible due to such factors as language barrier or altered mental status. In this instance, a low osmolar contrast should be utilized (*American College of Radiology Contrast Guidelines*. Chicago: American College of Radiology, 1990).
   C. **Compromised limb.** Patients with severe peripheral vascular disease appear to be at greater risk for significant local complications due to extravasation. This may also apply to patients with recent trauma and devitalized tissue near the injection site. Animal models demonstrate that low osmolar agents produce less local damage after subcutaneous injection (*Investigative Radiology* 25:678–685, 1990). Therefore, low osmolar agents are recommended for venography in these patients.

Chapter / 10
Myelography

Sven Ekholm

The very beginning of myelography started in 1921 with a paper by Jacobaeus about pneumomyelography. This technique had already been presaged 2 years earlier by Dandy in the first published paper about pneumoencephalography. In 1922, the first positive contrast medium for myelography was introduced, an oil-based agent called Lipiodol. It was not however, until the mid-1930s, before Lipiodol really achieved a widespread popularity as a myelographic agent, and by that time it had already been noticed that Lipiodol caused a lot of adverse reactions. The change in attitude was the result of two publications: one about the herniated disc syndrome, and the other about the use of Lipiodol in the diagnosis of herniated discs. In the early 1940s, Lipiodol was replaced by a newly developed, far less toxic oil-based competitor, Pantopaque, a contrast medium still in use to some extent. There are some definite limitations with all oil-based contrast media (CM), and major efforts were put into the development of water-soluble contrast medium from the very beginning. All of them had been too toxic for use above the lumbar region until the introduction of the first nonionic, water-soluble contrast medium, Amipaque. Amipaque was the result of a suggestion by Almen (*Journal of Theoretical Biology* 24:216–226, 1969), and it became the first real competitor to Pantopaque, since it had the advantage of being water soluble and could be used throughout the subarachnoid space (SAS). With increasing use it was found that Amipaque also had toxic properties, and it has now been replaced by several new, far less toxic, nonionic, water-soluble CM.

I. **Contrast media for myelography.** The CM in clinical use today almost exclusively belong to the water-soluble, nonionic group, but there may still be some institutions that continue to use

Pantopaque for certain indications. Gas as a true myelographic agent is outdated, but it is still used in combination with computed tomography (CT), i.e., to detect intracanalicular acoustic neurinomas, and that is why these last two agents are briefly discussed in this context.

A. **Negative contrast media.** The gases (air, oxygen, or nitrous oxide) have all the benefit of being nontoxic in the SAS, and they are readily absorbable following the examination. They allow visualization of the spinal cord, but they are not adequate to study fine structures, such as nerve roots and blood vessels. Moreover, when injected intrathecally in such a large amount as needed, they are associated with severe adverse effects due to a disturbed pressure balance, and this has given them a bad reputation among patients. When they are used in small amounts as in the examination of the acoustic canal with CT, this is really not a significant problem, but this technique is losing its importance with the increased availability of magnetic resonance imaging (MRI).

B. **Positive contrast media**
   1. **Oil-based contrast media.** The only oil-based contrast medium used today is Pantopaque, a well-tolerated contrast medium that displays a low incidence of serious reactions. On the one hand, it is easy to use, since it does not mix with the cerebrospinal fluid (CSF) or drain during the examination. On the other hand, the high density and lack of miscibility with CSF will result in a reduced visualization of small intraspinal structures. Properly used, Pantopaque has a low toxicity, but it is commonly accompanied with a modest pleocytosis with some elevation of total CSF protein. Serious reactions with neurologic symptoms are very rare. The more common complication is a chronic meningeal reaction, that in a small percentage of the patients may lead to severe adhesive arachnoiditis with disabling symptoms (*Journal of Neurology, Neurosurgery and Psychiatry* 41:97–107, 1978). **Arachnoiditis is said to be more common when there is a bloody spinal tap, and in those instances Pantopaque should be withheld.** This is also advisable because of an increased risk of intravasation and subsequent lung embolization in those cases. Pantopaque remains for an extensive time period (years) within the SAS, which may enhance the inflammatory reaction described above. For this reason it is generally agreed

that this contrast medium should be removed at the end of the examination.

All things considered, Pantopaque cannot be regarded as the most suitable contrast medium for myelography today. Moreover, the density and the properties of Pantopaque make it unusable in combination with CT. As a matter of fact, Pantopaque may even result in artefacts that can result in a suboptimal CT examination. The combination of myelography and postmyelography CT is very helpful when water-soluble CM are used. In some instances, e.g., to evaluate degenerative changes in the cervical spine, this combination of myelography and postmyelography CT will often be as informative as MRI. In MRI small amounts of residual Pantopaque can create diagnostic difficulties, sometimes misinterpreted as small metastatic lesions.

2. **Water-soluble contrast media.** The optimal myelographic CM should (1) be of a low viscocity, (2) be water-soluble, (3) be excreted at the end of the examination, (4) be nonionic, (5) be isoosmolar with CSF, and (6) be nontoxic.

This drug of choice has so far eluded us, but the gap between what's available and the ideal contrast medium has been narrowed over the years. From the first publication of a water-soluble contrast medium for lumbar myelography, Abrodil, to the introduction of the first contrast medium for general use in the SAS, Amipaque, have passed more than 40 years. What finally made the difference was the nonionic structure of Amipaque that has two significant impacts. There are no free ions that may interfere with neural function, and as a nonionic compound there will be a reduction in osmolality (isotonic with CSF in a concentration of about 170 mg I per mL). There were, however, some major drawbacks with Amipaque; it has been expensive and inconvenient to use (lyophilized powder to be dissolved before use). As mentioned earlier, it soon became apparent that this new contrast medium had some not insignificant adverse effects; most important were those related to neurotoxicity. However, in contrast to Pantopaque, there are no indications that Amipaque or its new successors will cause more inflammatory reactions or other adverse reactions when injected in conjunction with a bloody spinal tap.

Physiology. To understand better the origin of neurotoxicity with water-soluble myelographic CM and how neurotoxicity can be reduced, it is important to have a basic knowledge of central nervous system (CNS) physiology. The CNS is a unique and fragile system that needs a very stable environment to function properly. This is achieved through a variety of control mechanisms, most notably the blood-brain barrier (BBB), which acts as a CNS guard against many unwanted solutes from the blood, such as vascular CM (see Chapter 7). By means of a specific carrier system, certain important ions, amino acids, and D-glucose can be transported from the blood to the CSF and the neural tissue. This very delicate transport system guarantees the stability of the fluid composition of CSF and the neural tissue, since most of the other solutes in the blood will be kept out or enter very slowly (*Annals of Neurology* 13:469–484, 1983). Compounds injected directly into the CSF, such as myelographic CM, will bypass this barrier. Since there is no barrier between CSF and the extracellular fluid space, these agents will diffuse into the neural tissue and can come in direct contact with the neurons and may possibly interfere with their functions. Non-water-soluble CM, such as gases and oil-based CM, will remain in the CSF, and their neurotoxic potentials are thus very limited, compared with the water-soluble agents, but they may cause other reactions as mentioned above.

The factor most important to reduce the neural tissue concentration of CM and thus the toxicity problems is a continuous and good production of CSF. Most of the CSF is produced in the intraventricular choroid plexus, but a smaller portion is produced in the neural tissue as a result of normal metabolism. The metabolic fluid production, although small, is probably important in the discharge of metabolic slag products and will result in a sink effect that to some extent may counteract also the inward diffusion of CM. The water-soluble CM mix with CSF and are transported out of the SAS along with CSF. Consequently, any factor reducing CSF production or drainage will result in a delayed clearance of CM from the SAS, prolong the contact time, and increase the risk of neurotoxicity. Factors that may have a negative impact on CSF production are listed in Table 10.1.

The total CSF volume in the adult male is about 150–200 mL which is replaced about 3 times a day. This continuous production and pulsatile volumetric changes of the brain and the choroid plexus are the main factors transporting CSF out of the ventricular system. From the cisterna magna region, CSF has a

Table 10.1. Factors That May Reduce CSF Production

Dehydration
Rapid increase in serum osmolality
Intravascular CM
Alkalosis, especially respiratory
Acetazolamide (IV in animals)
Furosemide (IV in animals)

downward gravitational flow along the spinal cord to the lumbar region where some of it is drained. Some of the CSF will also pass around the brainstem and over the hemispheres to drain through the longitudinal sinus granulations. The exact relationship between lumbar and intracranial drainage of CM is not quite clear, but the major portion is probably drained closest to the site of examination. This is also supported by animal studies (*Acta Radiologica: Diagnosis* 26:331–336, 1985). Moreover, if the lumbar root of drainage had been negligible, the toxicity of the earlier, water-soluble, ionic CM had made them obsolete for lumbar myelography as well, since they otherwise had had to be transported upward along the spinal cord. As mentioned above, any factor influencing CSF drainage will have similar risk potentials as a decrease in CSF production. Some of the factors reducing CSF drainage are listed in Table 10.2.

II. **Symptomatology and origin of neurotoxicity.** The adverse reactions most often noted following myelography with non-ionic, water-soluble CM are minor, such as **headache, nausea,** and **vomiting.** These are symptoms also associated with a spinal tap alone and thus can be related, to some extent, to CSF leakage. The incidence following Amipaque myelography was, however, higher than that and lasted in some cases for a few days. It has also been shown that Amipaque in large doses can cause arachnoiditis in animals, and this has been verified in a few clinical cases as well (*Archives of Neurology* 37:588–589, 1980; *Neurosurgery* 17:467–468, 1985). The real drawback with Amipaque was related to its neurotoxic properties that became more and more obvious with increasing number of examinations. Many papers have been published in this field, most of which are case reports of various reactions following myelography, such as seizure, aphasia, and cortical blindness. The majority of these reactions are probably the result of a direct local toxic CM effect on the neurons themselves. Support for this can be found in several articles discussing case reports where CM for one reason or another have reached a high concentration

Table 10.2. Factors That May Reduce CSF Drainage

> Reduced CSF production (lower pressure)
> Increased serum osmolality (relative to CSF)
> Normal pressure hydrocephalus
> Arachnoiditis
> Diabetes

within a certain region of the brain, e.g., around the left temporal lobe, causing aphasia, or around the occipital lobes, with cortical blindness resulting (*Archives of Neurology* 42:24–25, 1985). All these neurologic complications have been transient but very dramatic when present. Some of the toxicity problems are probably related to osmotic effects. Maly and Fex demonstrated an increase of calcium and magnesium ions in rabbit CSF relative to sodium and potassium following the injection of hyperosmolar solutions of Amipaque and Omnipaque. In isotonic solution, a similar but less pronounced effect was seen with Amipaque, while isotonic Omnipaque was inert in this aspect (*Acta Radiologica: Diagnosis* 1:65–71, 1984). Such a relative increase in calcium and magnesium ions could be the origin of the depressive effect associated with hyperosmolality shown in other experiments (*Radiology* 145:379–382, 1982). The exact mechanism for such ionic shifts are not known but could be related to disturbances of the energy-dependent membrane transport mechanisms.

Hypertonicity has also been shown in vitro to cause a marked depression in neural tissue glucose metabolism measured as a reduced CO_2 production. In isotonic solutions, neither Omnipaque nor Isovue has shown such metabolic effects, but Amipaque does, and this is obviously a molecular effect (*Investigative Radiology* 19:574–577, 1984; *Investigative Radiology* 21:798–801, 1986). Exactly what it is in the Amipaque molecule that interferes with glucose metabolism has long been debated. The presence of 2-deoxy-D-glucose in one of the side chains led Bertoni to suggest and demonstrate in vitro a competition between Amipaque and hexokinase (*Annals of Neurology* 9:366–370, 1981). Such a competition between CM and hexokinase could not be seen with Omnipaque or Isovue (*Investigative Radiology* 22:137–140, 1987). Amipaque will reduce glucose uptake in neural tissue in vitro as well as in vivo (*Acta Radiologica* 31:209–212, 1989), but since Amipaque has never been shown to pass intracellularly, the toxic effects are most likely related to some transmitter function and/or disturbance of the cellular membrane itself. If the metabolic disturbance described above is the causative factor or if it is secondary to other effects is still too early to tell. Whatever the origin of this toxicity, the newer myelographic agents presently in use are far less toxic than Amipaque.

V. **Preparation and precautions for myelography.** The best preparation of a patient for myelography is to make the patient feel confident for the procedure and the radiologist performing the examination. To achieve this, the patient should always get a detailed and honest explanation by the radiologist about how the examination is done and what the patient can expect to experience. A confident patient is much easier to examine, and the quality of the examination will in most cases improve as a result of the subsequent cooperation from the patient. Moreover, if such a relationship is created between radiologist and patient, there is no need of sedatives in most cases involving adult patients. In spite of everything, there are always a few extremely anxious patients who will need some sedative support. In most instances, 5–10 mg of Valium are sufficient to achieve the intended result.

From what has been said earlier about CSF production (Table 10.1) and drainage (Table 10.2), one should keep in mind that the patient has to be well hydrated at the time of examination. Reduced serum osmolality will improve CSF production as well as drainage, while dehydration will result in a reduced CSF production and prolonged contact time between neurons and CM. **Consequently, the patient should have a liberal liquid diet until about 2 hours before myelography to reduce CM toxicity.** Solid foods should be withheld on the day of examination to reduce the risk of aspiration. Since such a delay in drainage can increase the incidence and severity of adverse reactions, one should always consider an IV drip infusion if the patient is unable to drink enough.

In the days of Amipaque myelography a lowered seizure threshold was generally regarded as a relative contraindication unless it was caused by medication that could be discontinued for at least 48 hours before the examination. When there was a definite need to use Amipaque for myelography in patients with epilepsy, it was recommended to premedicate with phenobarbital or other anticonvulsive drugs. Seizure and other severe adverse reactions are extremely rare with the new CM replacing Amipaque, which is why it is possible to examine also patients with known seizure disorder without any particular premedication. They should, however, continue with their usual anticonvulsive drugs, which can be supplemented with Valium if desired. The only contraindications for myelography left are the same as those for lumbar puncture and known allergy to the CM. When reduced drainage of CSF can be expected, e.g., in patients with chronic arachnoiditis, this is regarded as a relative contraindication. In those instances, it may be better to refer the patient for MRI primarily.

V. **CM injection and dose.** In most instances, it is possible to use a lumbar approach to examine all levels of the spinal canal. Also, examinations of the cervical region will generally result in a good CM concentration if the patient can hyperextend his or her neck and accept the rather quick head-downward tilt of the table. It is important to use lateral fluoroscopy and watch the neck region during the tilt to be able to catch the CM bolus when it approaches. In older patients and patients who are unable to hyperextend the neck, it may be necessary to use the lateral C1-2 approach, since it is easier in those instances to keep the CM in the region of interest. The recommendations from the manufacturers currently limit the use of CM to 3 gm of iodine in the adult patient. This is more than enough in most cases, but in some situations, e.g., in patients with a very large dural sac, it can be of help to use a somewhat larger volume and dose. A high-dose study of Omnipaque in which a maximum of 4.5 gm iodine was injected for lumbar myelography did not show any major adverse reactions. There was, however, an increase of minor reactions, such as headache, at the highest dose levels (*Radiology* 163:455–458, 1987). In children it has been recommended not to exceed 100 mg I per kg body weight. The use of body weight has been debated, since the size of the CNS is better related to age than weight. The volume and concentration to be used in the adult patient can be found in Table 10.3.

What should be stressed to improve quality is the importance of a good needle position before injection is started. During injection, the location of injected CM should be checked regularly to reduce the risk of inadvertent injection of large CM volumes outside the SAS. **It is also important to perform the injection slowly over 1–2 minutes to minimize turbulence and mixing with CSF, which will reduce the opacity and, as a consequence, the quality of the examination.**

VI. **Postmyelography measures.** Following the examination routine, bed rest has been customary in the past to reduce the risk of spinal tap headache. Comparative studies of patients kept

Table 10.3. Recommendations about Concentration and Volume of CM to Use in the Adult Patient

Level	Concentration (mg I per mL)	Volume (mL)	Maximum Recommended Volume (3 gm I)
Lumbar	180	12–14	17
Thoracic	240	12	12–13
Cervical			
Lumbar	240	10	12–13
C1-2	240	5–7	12–13

ambulant and those confined to their bed have not shown any significant difference, and in some cases the incidence of headache and nausea has been somewhat lower in the ambulant patient (*Radiology* 138:625–627, 1981; *Clinical Radiology* 34:325–326, 1983). **The most important factor to reduce minor adverse reactions with the new CM is probably to use thin needles for puncture, preferably not more than 22 gauge.** Bed rest seems not to be mandatory, and if the examination takes place in the morning, it should be possible also to send the patients back home at the end of the day if the postmyelography period has been uneventful.

II. **Indications for myelography.** Myelography has lost some of its importance today, first with the introduction of CT that allowed the diagnosis of herniated lumbar discs without intrathecal CM, and lately even more so with the general spreading and increase in the number of MRI units. As a result of the continued improvement in technique and quality of MRI examinations, many of the previous indications for myelography have been abandoned. In states where there is no or a very limited access to MRI, myelography is still the most important modality in spinal symptomatology, but everywhere else MRI is now becoming the primary tool in most of these cases. There are, however, still some indications left where myelography is preferable or necessary: (1) patients with solid metal rods where the rod will create severe artefacts in the area of interest; (2) patients that cannot cooperate to obtain diagnostic images; (3) negative MRI in cases of suspect spinal vascular malformation; (4) some cases of severe scoliosis, especially if diastematomyelia has to be ruled out; (5) some cases of degenerative disease where the surgeon may request a more detailed view of the bony changes (most often combined with a postmyelography CT); (6) since many MRI units have no on-call service, it may be necessary in cases of acute cord compression to use myelography preoperatively to identify the level of compression; and (7) suspicion of spinal stenosis that may need imaging with flexion and extension of the back. In summary, one probably has to accept that myelography in the future with continued refinement of the MRI technique will be replaced by this technique and eventually follow in the lead of pneumoencephalography. In the meantime, remember that the addition of postmyelography CT in many cases may shed new light on a question raised by the myelography. The combination of the two modalities may in some cases give an information that is at least as good as that of an MRI examination.

Chapter / 11
Arthrography

Albert Alexander
Daniel I. Rosenthal

The indications for arthrography have changed rapidly in recent years as a result of improvements in arthroscopy and magnetic resonance imaging. However, arthrography remains an excellent, safe technique for evaluation of intra-articular abnormalities of both native and replaced joints.

General indications include the evaluation of both intra-articular and juxta-articular abnormalities including ligamentous and tendinous tears, cartilage injuries, proliferative synovitis, masses and loose bodies, and implant loosening.

I. **Contraindications.** Although serious reactions to intra-articular contrast are extremely rare, a prior severe reaction to iodinated contrast agents is considered a relative contraindication to arthrography.

II. **Complications.** Arthrography is a safe procedure that has few complications when done properly. Possible complications include allergic reaction to the contrast, infection, sterile synovitis, and vasovagal reaction. Chemical synovitis was the most common complication in a survey done by Newberg et al. It was reported in 150 of 126,000 examination (*Radiology* 155:605–606, 1985). There were no deaths reported, and only three instances of sepsis were noted in this same group.

In our experience, by far the most common complication is postarthrography pain, which may represent a form of contrast synovitis (*AJR* 136:59–62, 1981). It is particularly troublesome in the shoulder and much less so in the hip and knee. Nonionic contrast agents appear to produce less pain (*Radiology* 154:339–341, 1985), and room air is less irritating than carbon dioxide (*AJR* 136:377–379, 1981) if double-contrast arthrography is performed. Postarthrography pain characteristically begins 4–6 hours after the injection, reaches peak intensity approximately

12 hours after the procedure, and then begins to abate. It is thus distinguishable from the pain of infection, which has a later onset and gradually increases in intensity. Nonsteroidal antiinflammatory agents are usually sufficient for management. In severe cases, postarthrography pain may persist for several days and require narcotics for relief.

III. **Sterile preparation and local anesthesia.** Individual practices may vary somewhat in details, but sterile preparation should resemble creation of a surgical field. We are especially fastidious when a joint implant has been done, since artificial joints are particularly vulnerable to infection, skin organisms such as *Staphylococcus epidermidis* are a common cause of infection, and if such organisms are found in a fluid sample, they cannot be assumed to be contaminants.

Whether to use local anesthetics is also a matter of individual discretion. In our opinion, if a local anesthetic is to be used, only a minimal skin weal is justified. We do not recommend the use of local anesthetic on small children, since it produces more agitation than it relieves pain.

   A. **Sterile preparation for nonprosthetic joints.** After the site of needle puncture has been determined, the skin is scrubbed 3 times each with a povidone-iodine (Betadine) solution followed by an alcohol-based surgical prep solution. Drapes are placed to form a sterile field, and local anesthesia is administered if desired.

   B. **Sterile preparation for prosthetic joints.** The same basic procedure as described above is used with some modifications. The modifications include: (1) all individuals in the room wear sterile masks and caps; (2) the person doing the procedure wears a sterile gown; (3) the surface of the fluoroscopic tower that faces the patient is covered with a sterile drape; (4) local anesthesia is not used because of its bacteriostatic properties and the possibility of obtaining false negative cultures from the joint fluid.

   C. **General principles**
      1. Preliminary films of the joint should be available prior to the injection of contrast. This is especially important in the evaluation of joint implants. To evaluate implant loosening, more than one view is required. For hip replacements, separate exposures made for the femur and acetabulum may be necessary because of different technical requirements.
      2. Joint fluid, if present, should be aspirated. We obtain

cultures of fluid recovered from joints with implants; otherwise, laboratory tests are done only as indicated.
3. Contrast is injected through sterile tubing so that fluoroscopy can be performed during the injection, if needed. When seen on the fluoroscope, contrast flows freely away from the needle tip if it is intra-articular.
4. When a single-contrast arthrogram is being done, it is wise to include some local anesthetic in the contrast mixture. Generally, a mixture containing at least 3 parts contrast to 1 part Xylocaine will not appear diluted on spot films. If tomography or computed tomography (CT) is to be done immediately after injection, a 1:1 mixture is more suitable.
5. For evaluation of loose bodies, single-contrast technique can be supplemented with tomography or CT.
6. If tomography is to be performed, 0.1–0.3 mL of 1:1000 epinephrine should be added to retard absorption.
7. We use a 25-gauge, 1½-inch-long needle for wrist, ankle, and elbow injections when there is no need to aspirate fluid. This is longer than the typical ⅝-inch needle but is widely available.

V. Arthrographic technique

A. Knee.
The patient is supine, with the knees slightly flexed over a sponge. Palpation of the midpoint of the patella can be used to identify the joint line. The skin is marked medial or lateral to the patella at this level, and the standard preparation is performed.

A 19-gauge needle (preferably with a short bevel) is introduced into the patellofemoral joint while the patella is retracted to the opposite side manually. A common error is anterior placement of the needle so that its tip strikes the patella. This causes pain for the patient. The approach should be almost horizontal. If the examiner advances the needle with one hand while slightly moving the patient's patella with the other, it is possible to determine when the needle is about to strike the patella and thus angle slightly deeper. Any fluid that is present should be aspirated, as a "dry" joint results in better quality images. If there is no fluid within the joint, the intra-articular position of the needle can be confirmed by attaching a syringe filled with air to the needle. If there is resistance to the injection of air, the needle is probably extra-articular and should be repositioned. Free flow of air into the joint can be felt by placing a hand over the suprapatellar recess of the joint. Following the injection of air the first syringe is removed, and a second syringe containing approximately 2–3 mL of iodinated contrast is connected. Before the contrast agent is injected, a few mL of air are aspirated to ensure that the needle tip remains intra-articular in location. The contrast is then injected, additional air is added, if necessary, to distend the joint moderately (as much as 35–50 mL may be needed), and the needle is removed. Flexing the knee once or twice is sufficient to distribute the air and contrast.

1. **Filming.** The patient is prone on the fluoroscopic table with the knee stabilized by one of the commercially available devices. A total of 9–12 fluoroscopic spot films are obtained of each meniscus. To ensure that the entire meniscus is examined, it is important to develop a consistent technique beginning posteriorly and working anteriorly or beginning anteriorly and working posteriorly. Generally, the meniscus should be outlined by both air and a thin coating of contrast. The knee is slightly rotated, and the patient is asked to turn if needed between exposures.

   TIP: Joint effusions are very common, and it is rarely cost-effective to send the fluid for laboratory studies unless there is a clinical suspicion of infection or crystal arthritis.

B. **Shoulder.** The patient is placed supine on the fluoroscopic table with the arm externally rotated. The midpoint of the glenohumeral joint is localized on the fluoroscope, and the skin is marked approximately 2 mm lateral to the apparent joint over the medial margin of the humeral head (Fig. 11.1). Standard skin preparation is performed. With the aid of intermittent fluoroscopy, a 20-gauge spinal needle is inserted vertically until the medial margin of the humeral head is contacted. Confirmation of intra-articular location can be obtained by injecting less than 1 mL of Xylocaine. If the needle is intra-articular, no resistance to injection is met. If resistance is present, the needle can be withdrawn approximately 3–4 cm, the needle hub can be displaced laterally (with the tip angled medially), and the needle can be reinserted while the bevel is kept directed laterally. The needle should slide over the humeral head into the joint.

   TIP: Another way of demonstrating intra-articular needle placement is to put the contrast mixture into a connecting

ARTHROGRAPHY 121

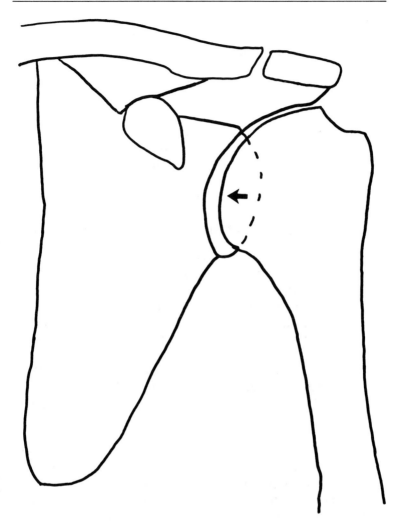

Figure 11.1. An anteroposterior view of the shoulder in line drawing. The arm is externally rotated, and the site of needle puncture is indicated by the *arrow*.

tube, making sure to keep a small amount of air at the needle end of the tubing. Under fluoroscopic observation a test injection is made. With intra-articular placement, if the syringe is depressed a tiny amount and released, air and contrast will rapidly flow away from the needle tip. If the needle is not within the joint, the air at the end of the tubing will be compressed and will "bounce back" when the pressure is released.

TIP: The most common error is superficial needle placement. Dense connective tissue anterior to the joint may feel surprisingly solid and be mistaken for bone. If the needle is on the humeral head in the correct position, it may be of help to lessen the amount of external rotation to loosen the anterior capsule.

Approximately 3 mL of iodinated contrast, 0.3 mL of 1:1000 epinephrine, and 9 mL of air are injected into the joint, and then the needle is removed. If CT is to follow the procedure, 0.3 mL of 1:1000 epinephrine is added to the mixture. We prefer single-contrast (9 mL contrast plus 3 mL Xylocaine) arthrography when there is a suspicion of adhesive capsulitis.

1. **Films obtained** are (1) anteroposterior (AP) internal rotation, (2) AP external rotation, and (3) axillary.

   If no rotator cuff tear is identified, the patient is asked to exercise the shoulder, and the AP views are repeated in the upright position. If a tear is identified, upright fluoroscopy of the shoulder may help to localize better the edges of the tear. CT (contiguous 5-mm sections through the joint) is added if there is joint instability or a history of prior dislocation.

   TIP: The acromion process slopes upward from back to front. In some cases, having the patient bend forward from the waist will help to project the acromion away from the rotator cuff.

   TIP: The upright position may be accompanied by vasovagal symptoms, especially in young men. Be alert to changes in color and mental status, and be prepared to lower the patient to a supine position.

C. **Hip.** The patient is positioned supine on the fluoroscopic table. The junction of the femoral neck and head is localized with fluoroscopy, and the skin is marked (Fig. 11.2). When a total hip replacement is present, the needle tip will be invisible because of dense metal. The site of puncture is usually at least 2 cm lateral to the femoral artery and below the inguinal ligament. If this is not the case with the patient supine, the patient can be turned into the anterior oblique position, with the hip to be studied elevated with a sponge. A pendulous abdomen can be kept out of harms way by tilting the head of the table down. Skin preparation depends whether a joint implant is present. A 20-gauge spinal needle is inserted vertically until the femur is encountered. In the

ARTHROGRAPHY 123

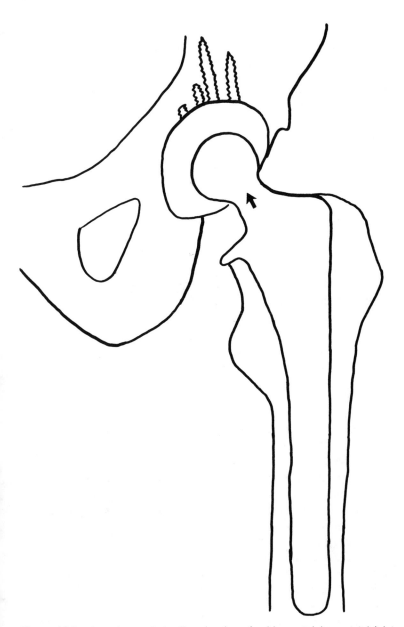

Figure 11.2. An anteroposterior line drawing of a hip containing a total joint replacement. The site of skin puncture is indicated by the *arrow*. The needle approach should be vertical (parallel to the X-ray beam). If necessary, the patient may be turned to ensure safe access to the joint (see text).

patient with a joint implant, the needle is inserted until metal is encountered. The intra-articular position of the needle can be confirmed by aspiration of fluid.

For patients with total joint implants, vigorous efforts to obtain fluid should be made before injection of contrast. Several maneuvers may be of help. If no fluid is initially aspirated, passive flexion of the hip with alternating internal and external rotation may be of help in expressing joint fluid. If fluid is still not obtained, redirect the needle so that the tip slides medially and/or laterally over the femoral head-neck junction to obtain intra-articular position. Aspirated fluid is then sent for Gram stain as well as culture and sensitivity. For those patients in whom fluid absolutely cannot be aspirated, needle position can be confirmed by injecting a tiny amount of contrast.

TIP: Patients with total hip replacements almost always have fluid that can be aspirated. A hemi-arthroplasty (such as an Austin-Moore implant) may result in a "dry" tap, especially when hip movement is restricted.

Approximately 5–20 mL of iodinated contrast are injected into the joint (depending on capacity). Injection is terminated when the joint appears distended or when discomfort is experienced by the patient.

1. **Films obtained (native hip)** are (1) AP and (2) frog leg lateral.
2. **Films obtained (hip replacement)** are (1) AP fluoroscopic spot film before injection to serve as a subtraction mask, (2) AP spot film after injection **with the patient and the fluoroscope in the same position,** (3) AP view of the hip, (4) frog leg view of the hip, (5) AP and frog leg lateral films after exercise, and (6) subtraction film.

   For a patient with total hip replacement, exercise will consist of walking from one end of the room to the other. The patient should be instructed to put weight on the affected side.

D. **Ankle.** With the patient supine, the midpoint of the tibiotalar joint is determined fluoroscopically, and a vertical line is drawn on the skin, marking the midline of the joint. The patient is then turned into the lateral position. A point is identified along the midline of the joint from which the needle can be advanced under the anterior tibial margin. Usually, this is approximately 1 cm caudal to the tibiotalar

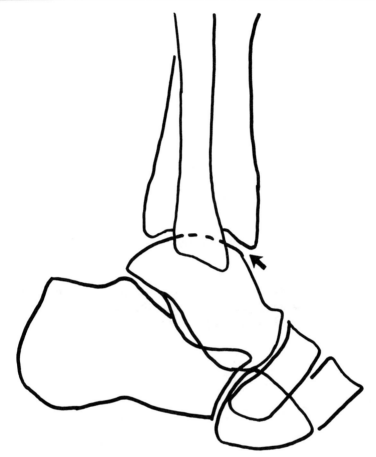

Figure 11.3. A lateral line drawing of the ankle, showing the direction of needle approach to the joint (*arrow*).

joint (Fig. 11.3). This is the site for needle puncture. If the dorsalis pedis artery is palpated at this site, the puncture can be made slightly medially or laterally. The standard preparation is performed. A 25-gauge, 1½-inch needle is then inserted with slight cranial angulation into the tibiotalar joint. Approximately 3–4 mL of a mixture containing iodinated contrast and Xylocaine are injected into the joint. Epinephrine (1:1000) at 0.3 mL can be added if tomography is to be performed.

1. **Films obtained** are (1) AP, (2) lateral, and (3) AP and lateral tomography as needed.

TIP: Lateral films obtained with the ankle extended and dorsiflexed will help distinguish synovial folds from intra-articular loose bodies.

E. **Elbow.** The patient is prone on the fluoroscopy table with the arm flexed 90° above the head to obtain a true lateral view of the joint. The joint between the radius and capitulum is identified (Fig. 11.4), the skin is marked, and the patient is prepped. A 25-gauge, 1½-inch needle is inserted vertically into the joint by using fluoroscopy as needed. A 2–3 mL mixture of iodinated contrast and Xylocaine is injected. Epinephrine is added if tomography is to be done. The needle is removed, and the elbow is flexed and extended briefly.

   1. **Films obtained** are (1) AP, (2) lateral, and (3) AP and lateral tomography as needed.

   TIP: Elbow arthrography is most often done to identify loose bodies. In our experience, this is best done with single-contrast technique—either all contrast or all air.

F. **Wrist.** The complete wrist arthrogram involves the injection of three separate compartments: the radiocarpal compartment, the midcarpal compartment, and the distal radioulnar

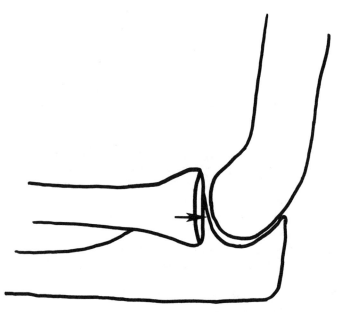

Figure 11.4. A line drawing of the elbow in lateral projection. The site of needle puncture between the radius and capitulum is marked by the *arrow*.

joint. There is some disagreement about the number of compartments to inject in order to identify significant pathology (*Program of the Ninetieth Annual Meeting*. Washington D.C.: The American Roentgen Ray Society, 1990, pp. 193–194). When a three-compartment arthrogram is done, the radiocarpal is injected first. The remaining two compartments are injected following approximately a 4-hour delay to allow resorption of contrast from the radiocarpal compartment. All needles are inserted from a dorsal approach. To inject the radiocarpal compartment, the wrist is flexed approximately 30° over a sponge, the midpoint of the radioscaphoid space is localized by using fluoroscopy, and the skin is marked (Fig. 11.5). It is advisable to keep the needle away from the scapholunate space. The standard preparation is performed, and a 25-gauge needle is inserted vertically into the joint space. A mixture of iodinated contrast and

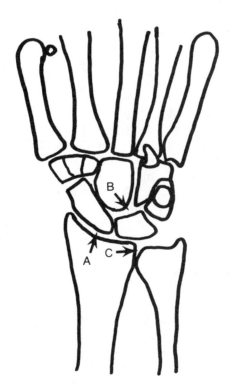

Figure 11.5. A posteroanterior line drawing of the wrist. *A*, the site of radiocarpal joint puncture; *B*, the site of midcarpal joint puncture; and *C*, distal radioulnar joint puncture.

Xylocaine is injected until slight discomfort is felt by the patient (usually 2–3 mL). This ensures adequate distension of the joint space. Fluoroscopy (preferably with digital subtraction capabilities) is performed during the injection to localize the site of any abnormal communications, and spot films are taken. The wrist is then exercised, and filming is done.

1. **Films obtained** are (1) posteroanterior (PA) neutral, (2) PA radial deviation, (3) PA ulnar deviation, (4) lateral, and (5) obliques.

   The second and third injections are performed in tandem. The hand is pronated and placed flat on the table. With use of fluoroscopy, the joint space where the lunate, triquetrum, capitate, and hamate converge is localized, and the skin is marked. The distal radioulnar joint is localized fluoroscopically, and the skin is marked. The forearm may be rotated slightly from the prone position for best demonstration of the cortical margins of the radius and ulnar at the sight of puncture. The standard preparation is performed for both areas. A 25-gauge needle is inserted vertically into the midcarpal compartment, and 2–4 mL of the contrast-Xylocaine mixture are injected. Again, fluoroscopy is performed during the injection. Fluoroscopic spot films are obtained in the PA projection, lateral projection, and oblique projection.

   Finally, a 25-gauge needle is inserted vertically into the distal radioulnar joint with use of intermittent fluoroscopy. A 2–3 mL mixture of contrast and Xylocaine is injected into the joint. PA and lateral films are obtained.

G. **Temporomandibular joint.** The temporomandibular joint is composed of two separate joint compartments completely divided by a meniscus (J. M. Taveras and J. T. Ferrucci, eds. *Radiology*. Philadelphia: J. B. Lippincott, 1986, Vol. 3, pp. 1–10). Displacement of the meniscus is a frequent cause of symptoms and can be most easily recognized by opacification of the lower joint space. Occasionally, the upper joint may be injected. If both can be opacified, meniscal position can still be determined. However, opacification of only the upper joint is difficult to interpret.

   For injection of the lower joint, the patient assumes the decubitus position, resting on one shoulder with the head tilted against the fluoroscopy table. In this position, the temporomandibular joint to be studied will be visible through the opposite cranium. The opening and closing of the mouth is observed fluoroscopically. With the patient's mouth closed, a point on the superoposterior surface of the mandibular condyle (at approximately the 1–2 o'clock position) is identified (Fig. 11.6). The skin is marked, and the patient is prepped. To avoid possible facial nerve block, deep anesthesia should not be used in order to avoid possible facial nerve block. A 23-gauge needle (preferably with a relatively blunt tip) is directed vertically onto the mandibular condyle. The patient is instructed to open his or her mouth partially. The examiner then gently "walks" the needle tip over the posterior margin of the mandibular condyle and advances it 5 mm into the joint space. Approximately 1–1.5 mL of a mixture of contrast, Xylocaine, and epinephrine are injected. (For this procedure, less than 0.1 mL of epinephrine is sufficient.) Intra-articular needle position is confirmed if contrast material flows rapidly away from the needle tip, over the condyle, filling the anterior recess. Extravasated contrast may roughly conform to the condylar contours posteriorly but will not flow anterior to the condyle. If the upper joint is entered inadvertently and the approach to the lower joint is obscured, the examination may have to be

Figure 11.6. A transcranial lateral view of the temporomandibular joint. The needle is advanced until it contacts the mandibular condyle at the site indicated by the *arrow*. The patient is then instructed to open his or her mouth slightly, and the needle is advanced into the lower joint.

repeated at another time. Fluoroscopy and spot filming are performed while the patient opens and closes the mouth.

TIP: A frequent cause of failure is excessively posterior needle position. If the patient opens the mouth too widely, it may be necessary to angle the needle anteriorly.

1. **Films obtained** are (1) fluoroscopic spot films in the open- and closed-mouth positions, and (2) lateral polytomography in the open- and closed-mouth positions.

Suggested Reading

Freiberger RH, Kaye JJ. Arthrography. New York: Appleton-Century-Crofts, 1979.

Goldman AB, Dines DM, Warren RF. Shoulder arthrography. Technique, diagnosis, and clinical correlation. Boston: Little, Brown, 1982.

Dalinka MK. Arthrography. New York: Springer Verlag, 1980.

Arnot RD, Horns JW, Gold RH, Blaschke DD. Clinical arthrography. Baltimore: Williams & Wilkins, 1981.

Chapter / 12
Hysterosalpingography

Amy S. Thurmond

I. **Indications** (Table 12.1). Hysterosalpingography helps to visualize the cervical canal, uterine cavity, and fallopian tubes and, therefore, is particularly useful in the diagnosis of women with infertility. Hysterosalpingography is also used to look for uterine anomalies as a cause of recurrent abortion or uterine masses as a cause of dysfunctional uterine bleeding. Hysterosalpingography can also be used to diagnose fistulas and in the past was used to diagnose ectopic pregnancy, intrauterine device location, and extent of uterine cancer (*Fertility and Sterility* 30:636, 1978).

II. **Limitations.** Some authors have recommended that hysteroscopy and laparoscopy be used instead of hysterosalpingography, since hysterosalpingography may fail to diagnose periadnexal adhesions to up to 30% of patients with this condition. Hysterosalpingography may also overdiagnose proximal tubal obstruction, with up to half of proximal tubal obstructions not confirmed at the time of surgery. Nonetheless, the majority of gynecologists believe that hysterosalpingography is a high-yield, low-risk test that should be employed in the initial evaluation of the infertile couple (*Fertility and Sterility* 40:139, 1983). Good technique and careful interpretation of the findings will maximize the benefits of hysterosalpingography. If there is suspicion of peritubal adhesions by history, and the hysterosalpingogram looks normal, laparoscopy may be considered. If proximal tubal obstruction is visualized despite adequate injection pressures and proper positioning of the patient (see Section IV, Technique), fallopian tube catheterization with selective salpingography may be considered prior to surgery to avoid false positive diagnoses (*AJR* 156:33, 1991).

Table 12.1. Hysterosalpingography

Primary indications
Infertility
Recurrent abortions
Contraindications
Active pelvic infection
Active vaginal bleeding
Recent[a] pregnancy
Recent[a] uterine curettage
Recent[a] uterine or tubal surgery
Possibility of pregnancy
Previous reaction to iodinated media

[a]Less than 6 weeks prior.

III. Contraindications (Table 12.1)

A. Active vaginal bleeding (to prevent the flushing of clots into the peritoneal cavity).

B. Active pelvic infection.

C. Recent pregnancy, uterine surgery, tubal surgery, or uterine curettage. One should wait 6 weeks before performing hysterosalpingography, to prevent venous intravasation of contrast medium.

D. Possibility of pregnancy. To avoid irradiating an early pregnancy the **ten-day rule** can be used. That is, one should not perform the procedure if the interval of time from the start of the last menses is greater than 10–12 days. Since menses usually start 14 days after ovulation, if the patient has cycles that are longer than 28 days, then the ten-day rule can be stretched to 13 or 14 days. If the patient has irregular cycles or absent menses, it is prudent to recommend a pregnancy test prior to performing the examination.

E. Previous reaction to iodinated contrast agent may be a contraindication, depending on the circumstances.

IV. Technique.
Since many of the complications thought to be contrast medium related can actually be prevented by good technique, a brief overview of proper technique is in order (*Fertility and Sterility* 40:139, 1983). The patient is placed in the dorsal lithotomy position, and the cervix is exposed with a speculum. The cervix and vagina are copiously swabbed with a cleansing solution such as Betadine. The hysterosalpingography cannula is placed. Once correct placement of the cannula is confirmed, the speculum should be removed. Leaving the speculum in is uncomfortable for the patient and results in suboptimal films. With fluoroscopic guidance, contrast agent at room temperature is slowly injected, usually 5–20 mL over 1

minute, and films are exposed. Injection of contrast agent is halted when adequate free spill into the peritoneal cavity is documented or when the patient complains of increased cramping, which usually occurs when the tubes are blocked.

A. Cannulas

1. An acorn-tip cannula in the external cervical os and a tenaculum on the external cervix can be used. This equipment is generally easy to use; the tenaculum, however, causes pain and bleeding. The discomfort from the tenaculum can be lessened by using a topical anesthetic (*Journal of Reproductive Medicine* 35:533, 1990).
2. An acorn-tip cannula in the external os with a cup over the external cervix to which a vacuum is slowly applied may be used. This avoids the use of a tenaculum and is the technique I prefer (*Radiology* 174:571, 1990) (Fig. 12.1).
3. A balloon catheter slowly inflated in the uterine cavity or endocervical canal is a simple device to use. To visualize all the anatomy, a tenaculum may have to be used to straighten the uterus. Also, if the balloon is inflated in the uterus, additional films with the balloon deflated or pulled down into the cervix should be obtained to visualize the lower uterine segment and cervix (Fig. 12.2).

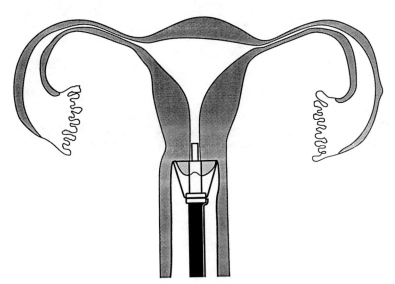

Figure 12.1. Diagram shows a vacuum cup-type hysterosalpingography device. The acorn tip is placed in the external os, the cup is slid over the outside of the cervix, and a vacuum is applied to the cup. This avoids the use of a tenaculum and allows visualization of the entire uterine cavity and cervical canal.

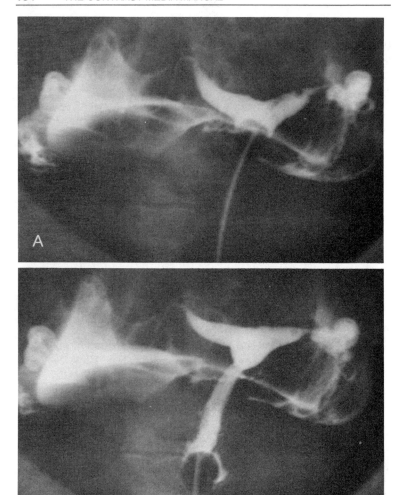

Figure 12.2. A normal hysterosalpingogram using a balloon-type device. (**A**) Initially, the balloon is inflated in the uterine cavity so that the tubes will be maximally filled with contrast medium. (**B**) To visualize the uterine cavity and cervical canal, the balloon must be deflated and withdrawn into the lower cervix and then reinflated if possible.

B. Films. In general, a kVp of 80 and phototiming give optimal films.
   1. An open film or scout film should be obtained to look for calcifications that may occur with ovarian dermoids or uterine fibroids. Contrast media from a previous oil hysterosalpingogram will also show up on the scout film.

2. An anteroposterior view when the tubes just start to fill will best visualize the uterine cavity and the interstitial and isthmic portions of the tubes.
3. Bilateral shallow oblique films will demonstrate the fallopian tubes and indicate whether the uterus is anteflexed or retroflexed.
4. A film showing the entire pelvis while traction is applied to the uterus will indicate the shape of the uterus and demonstrate whether the spill of contrast agent is free of loculated within the peritoneal cavity.
5. If there is a question of loculated contrast media or if there is a hydrosalpinx that may have a pinpoint opening, the patient should 360°, and another film showing the entire pelvis should be exposed.
6. If there is apparent proximal tubal obstruction, the patient should be placed prone, since sometimes anteriorly placed tubes will fill with this maneuver. Glucagon and other antispasmodics do not appear to help (*Investigative Radiology* 23:209, 1988).

V. **Complications**
  A. Discomfort or pain is commonly experienced by women undergoing hysterosalpingography (Tables 12.2 and 12.3) (*Fertility and Sterility* 38:1, 1982). Routine analgesia is not necessary. Reassurance and rapid and skillful completion of the examination are the best approach. Although the pain tolerance of the patient and the skill of the examiner are the main determinants of the level of discomfort, there are some data to suggest that spill into the peritoneal cavity is less painful with oil media than with water-soluble media including low osmolar or nonionic media (*Fertility and*

Table 12.2. History of Contrast Media Use for Hysterosalpingography[a]

Contrast Medium	Type	Approximate Date of Initial Use	Currently Used
Bismuth solution		1910	No
Lipiodol	Poppyseed oil	1925	No
Salpix	Water soluble	1953	No
Ethiodol	Poppyseed oil	1954	Yes
Sinografin	Water soluble	1959	Yes
Renografin-60	Water soluble		Yes
Conray-60	Water soluble	1966	Yes
Hexabrix-320	Water soluble	1982	Yes

[a]Data are from *Fertility and Sterility* 38:1, 1982.

Table 12.3. Common or Severe Complications of Hysterosalpingography

Complication	Incidence	Prevention
Pain	Varies widely	Reassuring and competent demeanor
		Skilled technique
		Topical anesthetic when tenaculum is used
		Slow inflation when balloon is used
		Slow application of vacuum when cervical cup is used
		Slow injection of contrast medium
Pelvic infection	Up to 3%	Doxycycline prophylaxis for high-risk patients (see text)
Intravasation	Up to 7%	Careful fluoroscopic monitoring
Severe bleeding	Very low	
Idiosyncratic or allergic reaction	Very low	(Be prepared to treat)
Vasovagal reaction	Very low	(Be prepared to treat)
Radiation of early pregnancy	Very low	Perform examination in follicular phase
		Insist on a negative pregnancy test in patients with absent or irregular menstrual cycles
		Question all patients about recent menstrual flow
Ovarian radiation exposure > 1 rad	Low	Judicious use of fluoroscopy and films

Sterility 38:629, 1982; *Radiology* 152:232, 1984; *Clinical Radiology* 36:533, 1985; *AJR* 138:559, 1982; Lindequist et al. International Congress of Radiology, Paris, 1989). Diatrizoate methylglucamine-iodipamide methylglucamine (Sinografin) and diatrizoate meglumine (Renografin) (personal observation) seem to cause more pain than the other water-soluble agents.

B. Pelvic infection is a serious complication of hysterosalpingography, causing tubal damage and, in the days before antibiotics, even death. In a private practice setting, the overall incidence of posthysterosalpingography pelvic infection was 1.4% (*American Journal of Obstetrics and Gynecology* 147:623, 1983). However, the incidence of posthysterosalpingography pelvic infection in the women with dilated tubes was 33%. Subsequently, when women with known dilated tubes were given prophylactic doxycycline, no postprocedure infections occurred (*American Journal of Obstetrics and Gynecology* 147:623, 1983). For this reason, if dilated tubes are noted at the time of hysterosalpingography, particularly if there are dilated tubes with free spill, doxycycline 200 mg PO should be administered before the patient leaves the department, and a prescription for 5 days doxycycline 100 mg PO bid given to the patient. If the tubes are not dilated, there is no evidence to suggest prophylactic antibiotics are necessary. However, all women should be warned about the symptoms of pelvic infection.

C. Mild vaginal bleeding is common after hysterosalpingography. Severe bleeding requiring curettage is unusual and is presumably related to underlying pathology such as endometrial polyps. If a tenaculum is used, cervical laceration can occur and may require administration of topical silver nitrate to stop the bleeding.

D. Lymphatic and venous intravasation occur in up to 7% of patients having hysterosalpingography. It can occur in normal patients but is usually related to tubal disease and/or obstruction, recent uterine surgery or curettage, uterine malformation or fibroid, misplacement of the uterine cannula, or excessive injection pressure or excessive quantity of contrast media (*Fertility and Sterility* 38:1, 1982). Early intravasation may be difficult to identify and may be confused with tubal filling. Intravasation is characterized by filling of parallel, usually beaded channels and a superior course (Fig. 12.3). Fallopian tubes are single channels and usually have a more lateral or inferior course. **When intravasation is recognized, injection should be stopped.** Intravasation of oil media should particularly be avoided, since it has been estimated that about 20% of patients with oil intravasation experience symptoms of oil emboli, including chest pain, cough, dyspnea, light-headedness, confusion, headache, coma, or even death (*Fertility and Sterility* 38:1, 1982; *Fertility and Sterility* 53:939, 1990). The more serious symptoms are associated with larger volume intravasation. As in lymphangiography, oil injection should not be performed in patients with a cardiac septal defect or severe pulmonary disease.

E. An allergic or idiosyncratic reaction related to the contrast medium can occur after hysterosalpingography, though the incidence is unknown and is presumably low. It is usually related to venous intravasation, with the usual manifestations. A delayed reaction consisting of urticaria and severe hypotension 1 hour after normal hysterosalpingography without intravasation was reported. The delayed reaction

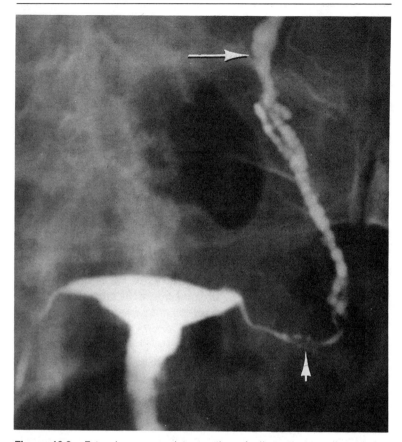

Figure 12.3. Extensive venous intravasation of oil contrast medium during hysterosalpingography. Lymphatic and venous filling often shows parallel channels (*short arrow*), and venous filling demonstrates a typical superior course (*long arrow*). When intravasation is noted, injection of contrast medium should be halted, particularly if oil is being used.

was presumably related to peritoneal absorption (*Fertility and Sterility* 54:535, 1990). To my knowledge, no deaths from idiosyncratic reactions to contrast media have been reported after hysterosalpingography.

F. Other side effects, such as vasovagal reactions and hyperventilation, occasionally occur.

G. Radiation exposure is a concern, since the women being examined are of reproductive age. The incidence of irradiating on early pregnancy is quite low when hysterosalpingography is routinely performed in the follicular phase of the cycle as described above. Although cases are few, there is nothing to suggest that inadvertent performance of hysterosalpingography in early pregnancy is harmful to the fetus (*Acta Radiologica: Diagnosis* 27:711, 1986). Radiation exposure to the ovaries is minimal and can be reduced by using good fluoroscopic technique and obtaining only the number of films necessary to make an accurate diagnosis.

H. Foreign body granulomas are a concern when oil contrast media are used, since oil is not readily reabsorbed from the tubes or the peritoneal cavity. There is no doubt that foreign body granulomas do occur after oil hysterosalpingography, though the incidence and the clinical significance are uncertain (*Fertility and Sterility* 38:1, 1982).

V. **Pregnancy rates after hysterosalpingography.** Pregnancy rates after hysterosalpingography have been a hotly debated issue for many years. Multiple retrospective or nonrandomized studies showed an increased pregnancy rate following hysterosalpingography, particularly when oil-based media were used (*Fertility and Sterility* 38:1, 1982). Recently, a randomized prospective study showed a significantly increased pregnancy rate after the use of oil medium than after the use of three water-soluble media including a nonionic agent (Lindequist et al. International Congress of Radiology, Paris, 1989). Even though all of these studies are subject to criticism, there probably is an unexplained therapeutic benefit of oil contrast media, particularly in patients with a normal uterus and fallopian tubes in whom the cause of infertility is unexplained.

VI. **Choice of contrast agent for hysterosalpingography.** The unique features of the commonly used contrast agents are summarized in Table 12.4. There is no clear-cut best choice, though strong proponents of both oil-soluble media and various water-soluble media exist. It is a personal choice on the part of the radiologist or the gynecologist and is often related to his or her training. In general, oil media result in higher contrast and sharper margins, though small intrauterine masses and the ampullary folds may be obscured (Fig. 12.4). Many gynecologists insist on the use of oil media in their patients because they believe it is therapeutic, while others forbid its use in their patients because they are worried about the consequences of oil emboli. A reasonable approach may be to use a water-soluble agent to perform the diagnostic examination. If this is normal, and there is no intravasation, oil may then be used to capitalize

Table 12.4. Unique Features of Various Contrast Media Used for Hysterosalpingography

Contrast Agent (Manufacturer)	Diagnostic Strengths	Pain with Peritoneal Spill	Bleeding Postprocedure	Potential Severe Complications	Cost	Postprocedure Conception
Water-soluble media						
Sinografin (ionic) (Squibb)	Folds visible	Increased				
Reno-60 (ionic) (Squibb)	Folds visible	?Increased				
Conray-60 (ionic) (Mallinckrodt)	Folds visible	Intermediate				
Hexabrix-320 (ionic, low osmolar) (Mallinckrodt)	Folds visible	Intermediate			High	
Omnipaque-240 (nonionic, low osmolar) (Winthrop)	Folds visible	Intermediate				
Oil-soluble medium						
Ethiodol (Sagave Labs)	Sharp margins	Decreased	Decreased	Oil emboli		Increased

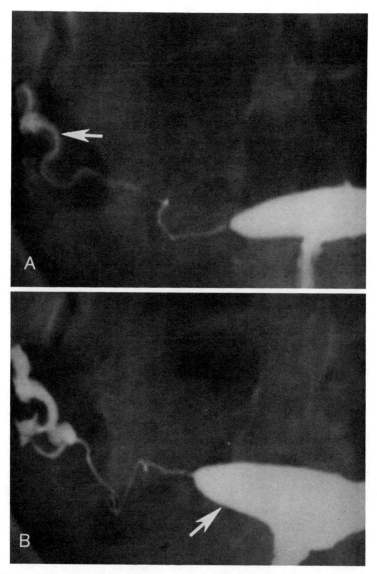

Figure 12.4. This infertile patient was examined by hysterosalpingography on two separate occasions. (**A**) Water-soluble contrast agent demonstrates the ampullary rugal folds of the right fallopian tube (*arrow*). (**B**) Oil contrast agent obscures the rugal folds; however, the margins of the uterus and fallopian tube are sharper (*arrow*).

on the possible therapeutic effect. As far as which water-soluble agent to use, it appears that Sinografin and Renografin may cause more pain with peritoneal spill than the others. Although data are limited, there is no evidence yet to suggest that the newer low osmolar or nonionic agents offer any advantages for hysterosalpingography. For these reasons, I use Conray-60.

Chapter / 13
Magnetic Resonance Contrast Agents

Val M. Runge

I. **Contrast agent design** (*Topics in Magnetic Resonance Imaging* 3(2):1, 1991)
   A. **Mechanism of enhancement.** Magnetic resonance imaging (MRI) is unique in that contrast is determined by many parameters (whereas with computed tomography (CT) there is only a single tissue parameter, X-ray beam attenuation) and the technique of measurement influences the observed contrast. Determinants of signal and contrast in MRI include spin density, flow (diffusion and perfusion), susceptibility, and relaxivity (T1 and T2). The bulk of research has focused on investigation of agents that alter tissue relaxation, and indeed the only approved agent, to date, gadopentetate dimeglumine (Gd-DTPA), falls in this category.
   1. Signal intensity in MRI is directly proportional to **spin density,** the number of hydrogen nuclei (protons). The body consists primarily of water protons, with little variability in concentration between tissues. To achieve contrast enhancement by an alteration in spin density, a marked change in tissue hydration is required. Despite the relative impracticality of this circumstance, two oral agents have received some attention, water and air.
   2. In MRI, the signal measured is the bulk magnetization in the plane perpendicular to the main static magnetic field. Motion during the process of measurement, including **diffusion** (the microscopic movement of bulk water) **and perfusion** (movement of blood in the microvasculature), induces a loss in "spin coherence" and thus observed signal. Neither property has been investigated with regards to development of contrast agents.
   3. The **susceptibility** of a substance is defined by the degree

to which it becomes magnetized in an external applied field. Four main categories exist: diamagnetic (small, negative), paramagnetic (positive), superparamagnetic (large, positive), and ferromagnetic (large, positive, with magnetization maintained when the field is removed). Study of substances containing paramagnetic ions dominates research in MRI contrast media, yet it is the property of relaxivity, not susceptibility, that is typically employed for enhancement. At high dose levels, paramagnetic substances might find application as susceptibility contrast agents for the measurement of tissue perfusion (*Radiology* 176:211, 1990).

MRI signal loss is seen at the interface between tissues with sufficiently different susceptibility. When small superparamagnetic particles are present in tissue, similar spin dephasing occurs but, in this instance, occupies a voxel and results in localized image intensity loss (negative enhancement). Particulate agents have been employed in this manner both intravenously (with phagocytosis by the reticuloendothelial system (RES)) for liver and spleen imaging and orally for opacification of the bowel.

4. **Relaxivity**-based contrast enhancement could rely on induced changes in either T1 or T2. The spin-lattice relaxation time (T1) and the spin-spin relaxation time (T2) constants characterize the rate at which the proton returns to ground state following excitation. The excess energy of the proton is transferred either to the surrounding environment of "lattice" (T1 relaxation) or to another proton (T2 relaxation). Most study of MRI contrast media has concentrated on paramagnetic agents that predominantly affect T1. These catalyze the longitudinal relaxation process and produce an increase in signal intensity on appropriate imaging techniques. However, all such agents affect both T1 and T2, although to differing degrees.

B. **Tissue specificity.** For an agent to be diagnostically useful, it must demonstrate a specific distribution or tissue localization. However, true targeting is rarely achieved. Utility of an agent will depend on the relative lesion-to-tissue distribution as well as the temporal time course. Paramagnetic metal ion chelates, such as gadopentetate dimeglumine, are extracellu-

lar in distribution, with efficacy in the central nervous system (CNS) due in a large part to the existence of the blood-brain barrier (BBB).
C. **Stability** of a contrast agent is essential, both in vitro (on the shelf) and in vivo. Chelates such as gadopentetate dimeglumine were designed to bind the gadolinium ion tightly (which otherwise would be quite toxic), preventing dissociation and enhancing rapid renal excretion.
D. Rapid **tissue clearance** of contrast media lowers the potential for chronic toxicity. Complete excretion by renal (as in the case of gadopentetate dimeglumine or other metal ion chelates) or hepatobiliary routes is desirable.
E. **Toxicity.** The efficacy of an agent to alter the parameter measured at a given concentration is of fundamental importance when dose and toxicity are considered. Both acute and chronic toxicity must be evaluated and are related to biodistribution, in vivo stability, and clearance.

II. **Imaging techniques for contrast agent detection**
A. **Spin density** is typically assessed clinically by using a spin-echo pulse sequence with as long a TR (>3000 msec) and as short a TE (<30 msec) as practical. This attempts to minimize contributions to signal intensity by T1 and T2 while maximizing contrast differences due to spin density. Acquisition times are long because of long TR.
B. **Diffusion** or microscopic motion results in signal loss within a voxel. This intravoxel incoherent motion has a negligible effect in most clinical imaging. The measurement of diffusion requires acquisition of multiple scans in which diffusion sensitivity has been changed, such as by (*a*) employing multiple echoes, (*b*) varying the gradient strength, (*c*) using additional gradients, and (*d*) changing gradient duration (*Magnetic Resonance Quarterly* 5(4):263, 1989). Diffusion imaging has not been employed, to date, for detection of contrast media but rather for its intrinsic value as an additional measurable tissue parameter.
C. **Susceptibility** contrast enhancement requires use of MRI techniques, such as gradient-echo imaging, that emphasize T2* (the total dephasing time, which is a combination of T2 effects and other dephasing processes such as susceptibility). Spin dephasing from T2 effects, field inhomogeneities, and susceptibility all contribute to T2*. Bulk susceptibility effects, such as that encountered at the interface between the

paranasal sinuses and soft tissue, lead to large image artefacts. This has limited, to date, the utility of susceptibility techniques in detection of contrast media effects.
D. **Relaxivity**-based imaging techniques are employed currently on a daily basis in clinical practice for observation of contrast enhancement with gadopentetate dimeglumine and other paramagnetic metal ion chelates. Most agents are designed to shorten T1 to a larger extent (without shortening T2). Most commonly, spin-echo methods are used to observe the effect of such agents, employing a short (≤ 600 msec) TR (increasing the T1 sensitivity of the scan) and short (≤ 25 msec) TE (decreasing the T2 sensitivity). Gradient-echo scans with short ($10 \leq TR \leq 200$ msec) TR, short (≤ 10 msec) TE, small flip angles, and spoiling of the transverse magnetization can also be employed in either two-dimensional (for rapid dynamic scanning in any organ or for breath-hold imaging in the abdomen) or three-dimensional (high-resolution imaging of a tissue volume) modes for detection of contrast enhancement. In both spin-echo and gradient-echo techniques, the observed signal intensity is dependent on both T1 and T2, regardless of the "weighting" of the technique. Relaxivity agents reduce both T1 and T2, with the former effect leading to an increase in signal intensity and with the latter leading to a decrease in signal intensity. Thus the relationship between contrast agent concentration and enhancement is not linear (Fig. 13.1). At concentrations of paramagnetic agents generally achieved clinically, the result is "positive" enhancement. Very high concentrations (such as are occasionally encountered in the bladder as a result of renal excretion) or the use of T2-weighted techniques can lead to "negative" enhancement (visualization of the contrast effect as a reduction in signal intensity). Implied in this discussion is the need for care in selection of imaging techniques and clinical interpretation thereof in visualization of contrast media such as gadopentetate dimeglumine.
E. **Dynamic imaging,** with either susceptibility or relaxivity contrast imaging techniques, can assess tissue perfusion by observation of the first pass of contrast media through the tissue of interest following bolus injection. Sequential acquisition of images on a time frame of 1–2 sec/image is required, which can be achieved by using gradient-echo scans sensitive to susceptibility effects or, alternatively, TurboFLASH techniques sensitive to T1 (relaxivity) effects.

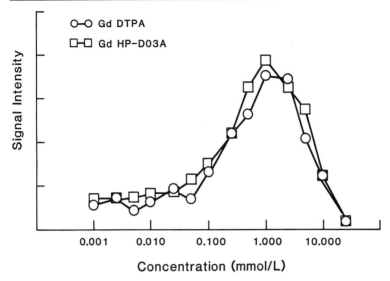

Figure 13.1. Signal enhancement of T1-weighted techniques is nonlinear with respect to concentration of contrast media (specifically a paramagnetic metal ion chelate). Results are from phantom studies at 1.0 T with spin-echo technique (TR/TE = 500/30). Little difference in enhancement characteristics is noted with the various gadolinium (*Gd*) chelates approved or near approval, to date.

II. Types of Agents

A. Intravenous. The compounds described below act as contrast agents in MRI by either enhancing T1 relaxivity and thus producing an increase in signal intensity on T1-weighted scans (Gd-DTPA, Gd-HP-DO3A, Gd-DTPA-BMA, Gd-DOTA, and Mn-DPDP) or by inducing highly localized susceptibility changes and thus producing a loss in signal intensity, which is best demonstrated on T2-weighted scans (AMI-25 and AMI-121).

1. **Extracellular distribution** (metal chelates). The T1 relaxivity and thus enhancement characteristics of these four agents are comparable, with few differences anticipated to impact clinical use (in utility of enhancement) other than that which might be achieved by increasing dose (Table 13.1).

    a. Gadopentetate dimeglumine (**Gd-DTPA** or Magnevist, Berlex Laboratories) was first approved for clinical use in the United States in 1988. The agent consists of the gadolinium ion (3+) chelated by DTPA (5−) (Fig. 13.2) and is formulated as the dimeglumine salt. In the United States, only a dose of 0.1 mmol/kg has been

Table 13.1. Contrast Agents in MRI

Agent	$LD_{50}{}^a$ (mmol/kg)	$T_{1/2}$ (Elimination) (minutes)	Osmolality (mOsm/kg)
Gd-DTPA	5.0–12.5	96	1960
Gd-HP-DO3A	10.7–13.6	94	630
Gd-DTPA-BMA	>11	78	789
Gd-DOTA	10.6–11.3	90	1400

^{a}Median lethal dose, i.e., the dose that is fatal to 50% of test animals.

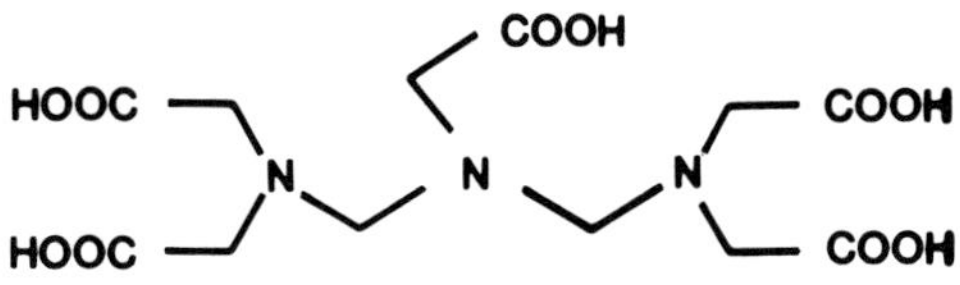

Figure 13.2. Chemical structure for DTPA.

evaluated (and is approved). In Germany, a dose of 0.2 mmol/kg is approved for the examination of intracranial neoplastic disease.

b. Gadoteridol (**Gd-HP-DO3A** or ProHance, Squibb Diagnostics) (*Radiology* 177:393, 1990) completed phase III trials in the United States in 1990. The agent consists of the gadolinium ion (3+) chelated by HP-DO3A (3−) (Fig. 13.3, a rigid macrocycle) and is thus nonionic (neutral). The formulation contains 0.5 M Gd-HP-DO3A, 0.25 mM Ca[Ca(HP-DO3A)]$_2$ (a scavenging agent for both free gadolinium ion and free chelate), and a buffer to maintain pH. Phase II trails evaluated dosages up to 0.3 mmol/kg, with an additional trial in progress to evaluate efficacy of high-dose administration in the CNS. No laboratory abnormalities attributable to contrast injection were observed in phase III trails, unlike those observed with gadopentetate dimeglumine (with which a transient elevation in serum iron and bilirubin can be observed, presumably on the basis of hemolysis). Stability relative to metal ion exchange is also improved, as compared with that for gadopentetate dimeglumine.

c. Gadodiamide (**Gd-DTPA-BMA**, Salutar) (*Radiology* 176:451, 1990) completed phase III trials in the United States in 1990. The agent consists of the gadolinium

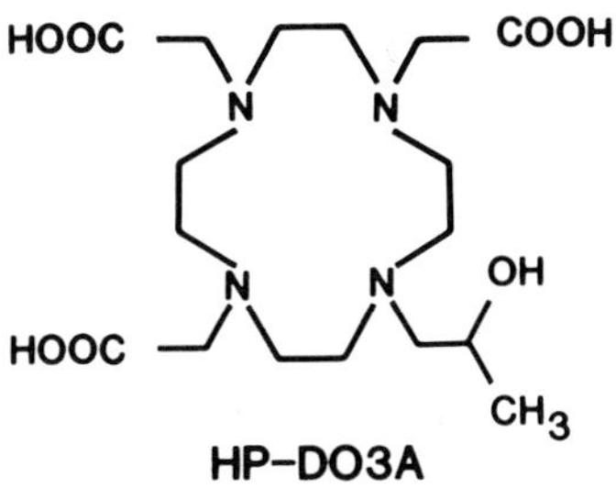

Figure 13.3. Chemical structure for HP-DO3A.

ion (3+) chelated by DTPA-BMA (3−) and is thus nonionic (neutral). Initial approval is being sought for a dose of 0.1 mmol/kg.

   d. **Gd-DOTA** (Dotarem, Laboratoire Guerbet) (*Investigative Radiology* 25:933, 1990) is approved in France and Portugal at a dose of 0.1 mmol/kg for clinical use in adults and children. The agent consists of the gadolinium ion (3+) chelated by DOTA (4−) (a rigid macrocycle) and is formulated as the meglumine salt.

2. **Hepatobiliary**
   a. **Mn-DPDP** (Salutar) has completed phase II trials in the United States. Enhancement of normal hepatic tissue can be observed postinjection on T1-weighted scans as a result of clearance by hepatocytes (with partial excretion of the agent in the bile). Detection of focal lesions should be improved due to their lack of enhancement relative to normal liver, analogous to that which occurs in nuclear medicine with ^{99m}Tc DISIDA. Initial trials indicate poor patient tolerance for this agent.

3. **Particulate**
   a. **AMI-25** (Advanced Magnetics), which consists of superparamagnetic iron oxide particles (*Radiology* 168:297, 1988), is scavenged by the RES following intravenous injection, permitting detection of lesions that displace RES cells, in particular, the Kupffer cells of the liver. Hypotension was observed in initial clinical trials, limiting investigation. Normal hepatic tissue decreases in signal intensity postcontrast, reducing the size threshold for lesion detection and permitting detection of a larger number of lesions.

B. **Oral**
  1. **Metal chelates** (positive enhancement)

    a. Gd-DTPA (Schering), formulated as 1.0 mM Gd-DTPA-dimeglumine with 15 gm mannitol per L water and administered at a dose of 10 mL/kg body weight, has completed clinical trials in Europe. Following oral administration, the bowel appears of homogeneous high-signal intensity on T1-weighted images. In this manner, intra-abdominal mass lesions and bowel wall thickening can be identified. On non-breath-hold sequences, images are degraded by peristalsis and resultant motion artefacts, with the latter accentuated by the high signal intensity of bowel contents. Mild diarrhea was reported in 4 of 20 patients (*AJR* 150:817, 1988).

2. **Magnetic iron oxide particles** (negative enhancement)

   a. Oral magnetic particles (Nycomed) consist of nonbiodegradable (nonabsorbed) spheres (3.7-micron diameter) containing crystalline magnetic iron oxide on a carrier matrix. These are administered at a concentration of 500 mg/L and a volume of 800 mL. Phase III trails are in progress in Europe. These iron particles cause a loss in signal intensity within bowel on the basis of superparamagnetic susceptibility effects. Thus, unlike orally administered gadopentetate dimeglumine, bowel peristalsis does not result in image degradation. However, the assessment of bowel wall thickening could be partially compromised due to effects on voxels surrounding that which contains the contrast agent.

   b. AMI-121 (Advanced Magnetics) consists of particles (0.2-micron diameter) containing iron oxide suspended in a low-viscosity flavored carrier (*Radiology* 175:695, 1990). On spin-echo sequences, a 900-mL oral dose containing either 100 or 175 mg/L was effective for bowel opacification. AMI-121 has completed phase III clinical trials in the United States. Imaging results are comparable to that of the Nycomed preparation. Mild diarrhea was noted in 5 of 15 volunteers.

IV. **Patient preparation** (IV gadopentetate dimeglumine). If contrast media are indicated, the risks and benefits of such should be discussed with the patient or guardian, and consent obtained. The process of informed consent should be documented in the dictation of the patient examination. Starting an IV line prior to placement of the patient in the scanner is recommended for time efficiency. The line should be checked immediately prior to contrast injection for patency. Gadopentetate dimeglumine is currently supplied at a concentration of 0.5 M and administered at a dose of 0.1 mmol/kg (equating to a dose of 0.2 mL/kg). The agent should be infused slowly and followed by a saline flush. The patient should be observed during and following injection for any possible reaction. After completion of the MRI examination, the patient should again be assessed, with the IV line discontinued if not further required. Administration of gadopentetate dimeglumine should be supervised by a physician, given the possibility of a severe anaplylactic reaction.

V. Indications

A. Head. IV contrast use of gadopentetate dimeglumine in the United States, to date, approximates 30% of all head examinations and continues to expand. It is difficult to identify prospectively the patients that will benefit diagnostically from contrast administration. Recommendations for use include all patients, except (potentially) children and young adults in which there is a low clinical suspicion of disease. A prospective clinical trial performed at a tertiary referral hospital (*Magnetic Resonance Imaging* 8:381, 1990) revealed "that addition of an enhanced exam will result in a change in diagnosis in 5%–8% of all cases" and provides "additional diagnostic information in 52%–69% of abnormal cases," as a result of both the presence of enhancement and the exclusion of significant pathology (the value of a negative enhanced examination).

Normal structures that have been shown to enhance include the choroid plexus, anterior pituitary, and infundibulum. Enhancement of both venous and arterial vessels is common but is influenced by flow rates, turbulence, timing, and other factors. Acquisition of both precontrast and postcontrast scans in the same plane is recommended for accurate diagnostic interpretation. (A more detailed discussion of indications is given in *Topics in Magnetic Resonance Imaging* 3(2):19, 1991).

1. Neoplasia

    a. For detection of small **extra-axial** lesions, such as acoustic neuromas and meningiomas, contrast administration is mandatory. Both entities can be missed on screening examinations without contrast because of their small size, isointensity with normal brain, and lack of secondary findings (such as mass effect or edema). Enhancement of these lesions occurs on the basis of intrinsic vascularity. Enhanced scans are also recommended for large lesions, for better assessment of lesion extent and, in postoperative cases, for assessment of residual and/or recurrent tumor. Extension of tumor along the dura can be seen in many meningiomas (this was a rare finding on CT scanning because of problems with bone artefact), providing for improved diagnostic specificity. Necrosis within large lesions is best assessed postcontrast. Care should be exercised in scan interpretation in order not to overlook diploic space lesions that, depending on size and vascularity, may enhance to isointensity with surrounding fat. Fat suppression sequences and high-dose contrast administration likely will have favorable impact on detection of diploic space lesions in the future. Meningeal disease, whether neoplastic or nonneoplastic, is difficult to assess without contrast administration. Caution is indicated, however, in interpretation of abnormalities involving the meninges, since enhancement may indicate little with respect to activity and nature of disease. Neoplasia, surgery, infection, and trauma may all produce meningeal disease as depicted by MRI.

b. Intra-axial lesions may display contrast enhancement on the basis of BBB disruption. With primary brain lesions, tumor bulk, necrosis, and tumor grade (low-grade astrocytomas do not typically enhance) at times are best assessed postcontrast. In screening for metastatic disease, contrast use is mandatory. In this instance, unenhanced scans may either reveal no lesions (due to the absence of sufficient edema or mass effect to allow detection), underestimate disease involvement, or simply be difficult to interpret. Preliminary studies indicate that high-dose administration (>0.1 mmol/kg, as evaluated in United States trials with gadoteridol) may be even more efficacious for lesion detection in metastatic disease.

Identification of tumor recurrence in the postoperative period is an additional indication for contrast enhancement. Caution is suggested in the interpretation of lesions postradiation, since radiation necrosis can present as an enhancing mass with surrounding edema (as on CT).

A small percentage of pituitary microadenomas are best detected postcontrast. Enhancement with large sellar and parasellar lesions can lead to better delineation of extent and cavernous sinus invasion.

2. **Infection.** Animal studies have established enhanced MRI as the modality of choice (versus CT and unenhanced MRI) for the detection of parenchymal infection. (*AJNR* 6:139, 1985). In both the cerebritis and capsule stages of evolution, enhancement can be noted due to BBB disruption. As might also be anticipated, contrast-enhanced MRI better delineates active inflammation of the meninges (tuberculous or bacterial), compared with unenhanced studies, and appears superior to CT. Unenhanced MRI studies are, however, more sensitive to ischemia, hemorrhage, and edema (*AJR* 154:809, 1990).
3. With the **demyelinating diseases** (e.g., multiple sclerosis), contrast administration provides for assessment of lesion activity (as well as disease progression or regression) and, in a minority of cases, identifies lesions not visible on unenhanced MRI.
4. **Infarction.** Four different temporal patterns of enhancement have been described with respect to early cerebral infarction (*Radiology* 177:627, 1990). Enhancement of vessels supplying the infarct ("intravascular enhancement," due to vascular engorgement and sluggish flow) can be seen 1–3 days postictus. At days 2–5, abnormal enhancement of the meninges adjacent to the infarct may be visualized. A transitional period, with vascular or meningeal enhancement combined with early parenchymal enhancement (due to BBB disruption), can be seen at days 3–6. In this prospective study, all patients (17 of 17) imaged at 7–14 days postictus demonstrated parenchymal enhancement. MRI perfusion studies, such as first-pass examination employing bolus contrast administration and subsecond dynamic imaging techniques (Turbo-FLASH), can be used to detect cerebral infarction in the first 24–48 hours postictus, prior to the onset of other detectable abnormalities.

In the subacute time period, enhancement both assists in differential diagnosis (by identification of BBB disruption) and permits, in a small number of cases, identification of a stroke that otherwise might not be recognized (due to the paucity of associated brain edema). The latter case is particularly true in the elderly population with underlying chronic deep white matter ischemia and gliosis.

5. **Arteriovenous abnormalities** (in particular dural arteriovenous malformations and venous angiomas) require contrast administration for detection in many instances. Magnetic resonance angiography techniques are also advocated (with and without enhancement). Both direct and indirect signs aid in interpretation, with the former due to enhancement of enlarged abnormal arterial and venous structures an the latter due to increased prominence of pulsation artefacts.

B. **Spine.** IV contrast use of gadopentetate dimeglumine for spine MRI in the United States, to date, is extensive, although less than that for head examinations, and continues to expand.

Normal structures that enhance include the venous plexus (both epidural and basivertebral), tectorial membrane and apical ligament, and nerve root ganglion. Enhancement in normal tissue occurs due to both intrinsic vascularity and the presence of fenestrated endothelium (in the dorsal root ganglion, muscle, and marrow), which allows the penetration of extracellular agents into the interstitial space. Enhancement in abnormal tissue occurs due to either an alteration in vascularity (extramedullary lesions) or breakdown of the blood-tissue barrier (intramedullary lesions). Acquisition of both precontrast and postcontrast scans in the same plane is recommended for accurate diagnostic interpretation. (A more detailed discussion of indications is given in *Topics in Magnetic Resonance Imaging* 3(2):41, 1991.)

1. In **degenerative disc disease,** enhancement both of venous plexus and scar tissue (whether occurring in the virgin spine or postoperatively in nature) provides additional valuable diagnostic information. Scar tissue is vascularized, with a large extracellular space accounting in part for its enhancement on MRI. In the virgin spine, prominent enhancement of both peridiscal scar and venous plexus (which may be dilated) improves conspicuity of disc herniation. Venous plexus enhancement alone improves delineation of disc from dural sac and depiction of foraminal disease. Enhancement of cord contusion, accompanying disc herniation, represents a caveat with respect to interpretation of intramedullary lesions (*AJNR*

10:1243, 1989). Type I end-plate changes can also be seen to commonly enhance.

In the postoperative lumbar spine, contrast enhancement can be used to differentiate between scar (which enhances in the immediate postinjection time period) and recurrent or residual disc (which does not). This forms the major indication of contrast use in spine imaging, to date, based on the number of patient examinations. Contrast-enhanced MRI is 96% accurate in differentiating scar from disc and is recommended in patients more than 6 weeks postoperatively (*AJNR* 11:771, 1990). Scar posterior to the thecal sac is different histologically from anterior scar, with enhancement decreasing after 4 months following surgery.

2. **Inflammatory lesions.** Epidural extension of disc infection and osteomyelitis are more readily identified and defined postcontrast. In the postoperative spine, however, it may be difficult to differentiate enhancement of an inflammatory mass from postsurgical changes. Meningitis with involvement of the subarachnoid space is best assessed postcontrast. Transverse myelitis has been reported to enhance.
3. **Neoplasia**
   a. Enhancement of **intramedullary** lesions may be used to direct surgical biopsy or can, on occasion, improve lesion detection. Astrocytomas tend to enhance in a patchy irregular pattern. Ependymomas and hemangioblastomas typically demonstrate intense enhancement. In the setting of a complex syrinx, contrast administration is mandatory for tumor recognition and delineation. Metastatic involvement of the cord itself can result from vascular or cerebrospinal fluid seeding, with small lesions best identified postcontrast. Acquisition of both sagittal and axial postcontrast images is recommended, in order to avoid missing lesions due to partial volume effects. There may be substantial cord edema accompanying metastatic lesions, best demonstrated on T2-weighted studies.
   b. **Intradural-extramedullary** neoplastic lesions, which include meningiomas, schwannomas, neurofibromas, ann drop metastases, typically all demonstrate marked enhancement. Small lesions may be difficult to detect on unenhanced MRI alone. CT myelography, on occa-

sion, is inferior to enhanced MRI for the detection of leptomeningeal metastastic disease.
  c. The primary role of IV contrast media in the **extradural** space is for delineation of soft tissue involvement. In metastatic disease confined to bone, postcontrast scans are often inferior to precontrast scans for lesion detection if imaging is not performed by utilizing fat suppression. On precontrast T1-weighted imaging, metastatic deposits are of typically low signal intensity, with enhancement decreasing their conspicuity relative to normal surrounding high-signal-intensity marrow. In the differentiation of osteoporotic and neoplastic compression fractures, as well as in the identification of bony metastatic disease in the elderly (with heterogeneous normal marrow), contrast enhancement can prove useful. Diagnostic utility may be improved further in these areas with the availability of high-dose contrast administration.
4. **Vascular disease.** In subacute spinal cord infarcts, enhancement may be noted. The slowly flowing venous component of arteriovenous malformations may also be seen to enhance, a significant observation is some cases for lesion detection.
5. **Demyelination.** Delayed contrast scans of the cord demonstrate lesion enhancement that correlates with clinical activity in multiple sclerosis (*AJNR* 10:1071, 1989). As symptomatology decreases, the magnitude of enhancement lessens. Delayed scans (45–60 minutes postinjection) are, however, required for detection of enhancement, with immediate postcontrast scans often revealing little change.
6. **Congenital lesions.** In neurofibromatosis, contrast enhancement may be of benefit for definition of plexiform neurofibromas and with occasional intramedullary cord tumors.

C. **Body.** Progress in development of applications for IV gadopentetate dimeglumine in the body (outside the CNS) has been slow, reflecting to some extent the evolution of MRI itself. Preliminary results support several indications, of particular impact in the head and neck and in the musculoskeletal system, since examination of these types constitute 20% or more of all cases at some sites. (For a more detailed discussion of indications, the reader is referred to *Topics in Magnetic Resonance Imaging* 3(2):74, 1991.)

1. In the **head and neck,** contrast administration has been found to improve lesion visibility, particularly with "tumors of the nasal cavity and paranasal sinuses and tumors having perineural or intracranial extension" (*Radiology* 172:165, 1989). In view of experience in other anatomical regions, enhanced scans will likely also be of benefit in infection. Use of fat suppression imaging techniques, only recently available, should impact favorably on contrast use in the head and neck, permitting better discrimination of enhancement.
2. In **bronchogenic carcinoma,** there is little indication that contrast media will be of utility. Neither CT nor MRI can accurately stage bronchogenic carcinoma, and both display poor sensitivity and specificity for nodal metastases and chest wall invasion.
3. **Acute myocardial infarction** can be detected with gadopentetate dimeglumine (*Clinical Cardiology* 9:527, 1968). There is delayed washout of contrast media in damaged compared with normal myocardium, resulting in enhancement of the zone of injury—typically during the first 2 weeks following acute infarction. Enhancement may be diffuse, subendocardial, inhomogeneous, or with a central nonenhancing core of tissue (presumed to represent a central nonperfused zone). No enhancement is observed in chronic myocardial infarction.
4. In **breast** imaging, malignancy no larger in size than the slice thickness can be excluded on the basis of no enhancement (S. H. Heywang-Kobrunner, ed. *Contrast Enhanced MRI of the Breast.* Karger, 1990). Focal enhancement can occur with malignancy, benign tumors, and (rarely) proliferative dysplasia. Diffuse enhancement is nonspecific (but most often represents proliferative dysplasia). MRI is of value in differentiating carcinoma from dysplasia and postoperative scar (with >6 months postsurgery or >18 months postradiation) and in determining lesion extent in mammographically dense breast.
5. In the **liver and spleen,** gadopentetate dimeglumine can reveal some metastatic lesions that otherwise might be poorly visualized. Many metastases become isointense with liver postcontrast as a result of diffusion of contrast into the tumor during the time required for data acquisition (typically 5–10 minutes). Breath-hold fast imaging techniques (currently under development) combined with

bolus contrast injection have potential for improved lesion detection. In a limited clinical series (*Radiology* 171:339, 1989), hepatocellular carcinoma could be differentiated from hemangioma on the basis of delayed enhancement, with such being absent or minimal in the former and moderate to marked in the latter. If safe formulation of magnetic iron oxide particles can be achieved and these become Food and Drug Administration (FDA) approved, improved detection of parenchymal lesions and, in particular, metastatic disease should be possible postcontrast.

6. No **bowel** formulations of contrast media are currently approved by the FDA. Distension of the stomach and rectosigmoid by air can be used to improve recognition of mass lesions and wall thickening. Likely to receive approval in the near future are oral gadolinium chelate formulations (which provide positive bowel contrast) and magnetic iron particles (which provide negative bowel contrast). Indications for bowel opacification include identification of the pancreatic head, differentiation of soft tissue masses from bowel, and improved definition of bowel wall thickening (V. M. Runge, ed., *Enhanced Magnetic Resonance Imaging*. St. Louis: C. V. Mosby, 1989, p. 301).

7. In the **kidney,** gadopentetate dimeglumine can be used both for improved detection of mass lesions (*Radiology* 176:333, 1990) and for assessment of renal function (*Radiology* 170:713, 1989). Renal masses may be isointense on nonenhanced imaging (both T1 and T2), requiring anatomic deformity for diagnosis. Lesion definition is improved after use of gadopentetate dimeglumine, particularly when dynamic imaging is performed. Contrast enhancement may also be of help in demonstrating extrarenal masses. Good correlation exists between ^{131}I-hippurate nuclear medicine studies and contrast-enhanced dynamic MRI for assessment of renal function. Differential temporal enhancement of normal renal cortex and medulla can be observed.

8. In the **adrenal gland,** differentiation between adenomas (moderate enhancement) and malignant lesions (marked enhancement with slower washout) is aided by contrast use. However, definitive diagnosis with regards to tissue histology cannot be achieved. Unenhanced T2-weighted imaging alone is able to achieve differentiation in 70–75% of cases.

9. In the **pelvis,** contrast enhancement can be of benefit in the assessment of endometrial, cervical, and vaginal carcinoma. Differentiation of the zonal anatomy of the uterus can be achieved on T1-weighted images postcontrast, with the junctional zone and cervical tissue remaining of lower signal intensity. Endometrial carcinoma enhances to a degree similar to the myometrium but less than the endometrium. Postcontrast studies improve differentiation of necrosis and retained secretions.

10. In the **musculoskeletal system** (*Magnetic Resonance Quarterly* 6(2):136, 1990), IV contrast injection is of value in neoplastic disease, osteonecrosis, and infection. Differentiation of tumor necrosis and delineation of tumor extent can be improved postcontrast. In healing osteonecrosis, reparative mesenchymal tissue enhances, permitting delineation of perfused viable repair tissue. Contrast enhancement can call attention to soft tissue infection, in addition to improving delineation of such. Intra-articular injection, which is in the early stages of investigation, may demonstrate improved delineation of traumatic injuries.

VI. **Contraindications** (Table 13.2) to IV contrast injection include **pregnancy, lactation, and renal failure** (V. M. Runge, ed. *Clinical Magnetic Resonance Imaging,* Philadelphia: J. B. Lippincott, 1990, p. 504). FDA approval at present is only for IV gadopentetate dimeglumine at a dose of 0.1 mmol/kg in the head and spine, in patients of 2 years of age or older. Gadopentetate dimeglumine is known to cross the placenta and to be excreted in breast milk. Elimination of this agent is primarily by glomerular filtration, thus caution must be exercised in patients with limited renal function. Chronic toxicity due to metal ion

Table 13.2. Contraindications and Relative Contraindications to IV Contrast Injection

Contraindications
  Pregnancy
  Lactation
  Renal failure

Relative contraindications
  Hemolytic anemia
  Wilson's disease
  History of anaphylactic reaction to iodinated agents

retention would be the clinical concern. Definitive studies in patients on dialysis have yet to be performed. In the infant, glomerular filtration is not fully developed, raising questions with respect to possible toxicity of the agent in this population, particularly since no systematic evaluation of the agent has been performed in children less than 2 years of age.

A. **Relative contraindications** (Table 13.2) include **hemolytic anemia, Wilson's disease,** and a **history of anaphylactic reaction to iodinated agents.** Gadopentetate dimeglumine administration can cause mild transient hemolysis (presumably on the basis of osmolality), with elevation in serum iron and bilirubin in some patients. Under certain in vitro conditions, copper and zinc ions can exchange for gadolinium in the chelate, with the implications of this clinically not documented. Anaphylactic reactions to gadopentetate dimeglumine have been documented (*Magnetic Resonance Imaging* 8:817, 1990), although they have been rare. Treatment of reactions to contrast media (and precautions prior to administration) should follow the guidelines established for iodinated media.

VII. **Complications** encountered with IV gadopentetate dimeglumine administration are few. These are best documented, to date, by a study of 1068 adult patients in the United States (*Radiology* 174:17, 1990). The most commonly reported treatment-related reactions included headache (3.6%), injection site coldness (3.6%), and nausea (1.5%). Vomiting was reported in 0.6% and rash on 0.3%. Blood and urine evaluation revealed only a "transient, asymptomatic rise in serum iron and bilirubin levels in some patients," thought to be a consequence of slight hemolysis. This phenomenon is consistently observed at a dose of 0.25 mmol/kg, 2½ times that employed clinically, to date, in the United States (*Investigative Radiology* 23(Suppl. 1):S275, 1988). As previously noted, there are reports of severe anaphylactic type reactions. In one documented case, within 15 minutes of contrast administration, the patient developed severe shortness of breath (accompanied by facial swelling and pruritus). There was prompt response to epinephrine administration.

Chapter / 14
Mechanisms of Contrast Media Reactions
Implications for Avoidance and Treatment Based on Hypothesis of Causation

Harry W. Fischer / Part I, Anthony F. Lalli / Part II
Elliott C. Lasser / Part III

Part I

I. **General considerations.** The ionic, water-soluble triiodinated contrast media (high-osmolality contrast media HOCM) employed for intravenous urography, angiography, and enhanced computed tomography (CT) scanning and, to a much lesser extent, for arthrography, gastrointestinal examination, and direct cholangiography and pancreatography have had extensive usage since 1952. During this time, about 5% of all patients receiving the ionic contrast media experienced some kind of reaction in addition to the sensation of warmth that most patients feel to some degree. About 0.05% (1 in 2000) had a reaction requiring hospitalization for treatment, and about 0.0025% (1 in 40,000) reactions were fatal. A higher incidence of reactions, with 1 fatality/10,000 examinations, was noted for the intravenous cholangiographic examinations. Reactions of any kind or severity are obviously not desirable, but fortunately most reactions are minor and self-limited and have no residual effect. Yet radiologists would like to have their patients freed from even minimal risk.

Since 1969 with the introduction and wider employment of the low-osmolality contrast media (LOCM), particularly with the second-generation agents (since the original first-generation agent, metrizamide, was little used intravenously), it has been increasingly apparent that the incidence of adverse reactions has been lowered. Evidence of this is reviewed first, and then the reasons why this is occurring are considered in light of mechanisms proposed to explain causation of adverse reactions.

II. **Studies comparing high osmolar to low osmolar contrast**

media. The first report based on large numbers of patients was that of Schrott in 1986 (*Fortschritte der Medizin* 104:153–156, 1986). Instead of a 5% incidence of all reactions, a 2.1% incidence of all reactions was noted in 50,660 intravenous urography examinations with the low-osmolality nonionic medium, iohexol, despite many of the patients being of high risk. The number of patients requiring hospitalization was reduced to one fourth of what had been experienced with ionic media.

Palmer (*Australasian Radiology* 32:426–428, 1988) reported on a large series of patients, a total of 109,546, from Australia on whom both ionics and nonionics were used. Low-risk patients had a higher incidence of adverse reactions with ionic high-osmolality media than did high-risk patients receiving nonionic low-osmolality media (Table 14.1). Reactions of all categories were reduced to one third by use of the nonionic, iopamidol.

In Austria in 1987, Hruby (*Rontgenblatter* 40:73–77, 1987) reported on over 50,000 patients who received high-osmolality ionic media or low osmolar nonionic media. A 6.98% incidence of adverse reactions for patients receiving ionics and a 0.07% incidence for patients receiving nonionics were reported.

Wolf (*AJR* 152:939–944, 1989) reported on a study comparing 6006 patients who received ionic contrast media with 7170 patients who received nonionic media (Table 14.2). This study showed also the pronounced reduction in reactions experienced by the patients receiving the nonionics.

The largest study was that of Katayama (*Radiology* 175:621–628, 1990) who reported on a total of 337,647 patients, almost equally divided between those receiving ionic and those receiving nonionic contrast media for intravenous urography, CT examinations, and intravenous digital subtraction angiography. The incidence of all reactions was 12.66% with ionic media, but only 3.13% with the nonionic media. A similar reduction was noted for the severe reactions, 0.22–0.04%, and for the very

Table 14.1. Reactions Reported in Australian Study by Palmer in High-risk and Low-risk Patients Receiving Ionic or Nonionic Intravenous Contrast Media[a]

Reaction	Mild (%)	Moderate (%)	Severe (%)	All (%)
High risk—ionic	7.2	2.7	0.36	10.3
Low risk—ionic	3.2	0.3	0.09	3.6
High risk—nonionic	1.1	0.1	0.03	1.3
Low risk—nonionic	0.9	0.09	0	1.0

[a]79,278 patients received ionic media; 30,268 received nonionic media.

Table 14.2. Reactions Reported by Wolf in Patients Receiving Ionic Contrast Media Compared with Patients Receiving Nonionic Media

Reaction	Ionic	Nonionic
Mild	2.5	0.58
Moderate	1.2	0.11
Severe	0.4	0.0
Requiring treatment	1.3	0.2
All	4.17	0.69

severe reactions, 0.04–0.0%. For those patients who had a history of adverse drug reaction on previous exposure to contrast media, 44.04% reacted again to ionic media, while the reaction rate in those patients receiving nonionic media was 11.24%. In this group of prior reactors, severe reactions were reduced from 0.73% for those patients receiving ionics to 0.18% for those receiving nonionics. From previous studies, patients with a history of allergy who are known to react with almost twice the frequency of patients in general were similarly found to react to ionics and nonionics with this frequency in the Japanese study, but the incidence of reactions was reduced from 23.35% for those receiving ionics to 6.85% for those receiving nonionics.

In the most recent study (G. L. Wolf et al. *Investigative Radiology* 25:S20–S21, 1990) in which a nonionic contrast medium was compared with ionic media alone and ionic media plus oral methylprednisolone, the nonionic agents were found to reduce total adverse drug reactions significantly, compared with ionics, and to reduce mild, moderate, and severe reactions significantly (Table 14.3). Steroid premedication provided some protection, but the nonionic, iohexol, significantly reduced mild, moderate, and severe reactions compared with those experienced in patients who were premedicated before receiving ionic media.

None of these studies were ideal in design. However, the evidence is overwhelming for the great reduction of adverse reactions by the use of low osmolar nonionic media. All six studies are insufficient in number of patients to determine if the incidence of fatal reactions is reduced through the use of the new media. **However, since the nonionic media greatly reduce the incidence of severe and very severe reactions and the difference between severe and very severe reactions and fatal reactions may, in large part, be the speed and skill with which a reaction is treated, the reduction in severe and very severe reactions should indicate fatal reactions will be reduced likewise.**

Table 14.3. Reactions in Three Contrast Media Regimens Reported by Wolf

Reaction	Ionic Agents[a]	Diatrizoate and Steroid[a]	Nonionic Iohexol[a]
Mild	2.9	2.9	0.44
Moderate	1.2	0.9	0.14
Severe	0.32	0.25	0.01
Requiring treatment	1.2	1.7	0.24
All	4.4	4.0	0.59

[a]Percentages are as a percent of all patients in the regimen.

III. **Why reactions are reduced with the use of nonionics.** The following gives some rationale for the reduction in reactions experienced with the use of nonionic contrast media. The mechanisms that have been proposed for causation of contrast media reactions are listed alphabetically in Table 14.4. Although evidence for each of these mechanisms has been considered, there is no single mechanism agreed on about why adverse reactions occur.

Clearly, the low osmolar contrast agents, particularly the nonionic low osmolar agents, produce **less alterations in the hemodynamic state of the body.** Ample numbers of studies of the pulmonary and cardiac circulation and the general circulation attest to this, in major part because of less pronounced fluid shifts that occur as the contrast medium courses through the circulation. Along with a lesser effect on cardiac contractility and electrical conduction, there is a lesser tendency to initiate cardiovascular reflexes that may be a factor in severe reactions characterized by cardiovascular collapse.

Low osmolar media have less effect on the **blood-brain barrier,** mainly demonstrated in cerebral angiography. If contrast media following intravenous injection exert their effects through changes in the blood-brain barrier, this may be a reason why fewer reactions are seen with LOCM. The lesser effect of the LOCM on the **microcirculation** has also been established. The lesser **red cell deformation and aggregation** seen with the nonionics reduced the likelihood of any adverse reaction being initiated through this mechanism.

Adverse reactions in which **bronchospasm** is a significant component are expected to be less frequent, since the nonionic media, iopamidol and iohexol, produce much less bronchospasm than are produced by an ionic HOCM. Whatever part the **inhibition of cholinesterase** by contrast media plays in the

Table 14.4. Mechanisms That Have Been Considered for the Cause of Adverse Reactions to Contrast Media

Antigen-antibody
Anxiety and emotions
Blood-brain barrier penetration
Blood changes
Bronchospasm and pulmonary edema
Calcium-sodium changes
Cholinesterase inhibition
Complement activation
Drug synergism
Hemodynamic and heart changes
Histamine release
Injection of impurities
Protein binding and enzyme interface

induction of adverse reactions should also be reduced, for the two nonionic media, iopamidol and iohexol, have been found to inhibit cholinesterase less than did an ionic medium.

The role of contrast media acting through **release of vasoactive substances** intravascularly to produce adverse reactions has received attention. In this regard, nonionics have a lesser effect on the **endothelium,** i.e., a less injurious effect that would lead to a less likely release of **vasoactive initiating substances from endothelium.** The new nonionics, iohexol and iopamidol, are weak **protein binders** and tend to **activate complement** less than do more avid protein binders. **Histamine release** is less with the nonionics than with the ionics, another possible factor in reduction of adverse reactions, although previously there has been no close correlation found between histamine levels and occurrence of reactions.

The low osmolality nonionic, iopamidol, and the low osmolar ionic dimer, ioxaglic acid, did not produce lowering of the **ionized calcium** in the blood after intravenous doses in clinical urography as did high osmolar ionic agents. The **binding of calcium** has not been shown to be related to the occurrence of reactions during intravenous use but may be important in production of ventricular fibrillation in coronary arteriography.

Of the possible mechanisms that may be the cause of adverse reactions, it is seen that the nonionics are less likely to initiate the mechanism. Of the several possible mechanisms, I believe the reduction of hemodynamic changes to be most important in minimizing adverse reactions.

Part II

I. Contrast media are associated with a **range of reactions** from minor to severe, including death (Table 14.5).
   A. **Minor reactions** with ionic media occur in 5% of intravenous injections and include nausea, heat or pain, and urticaria.
   B. **Intermediate reactions** with ionic media occur in 0.022% of intravenous use and include severe emesis, dyspnea, hypotension, and chest pain.
   C. **Severe reactions** with ionic media occur in 0.0025% of intravenous injections and consist primarily of shock, pulmonary edema, cardiac arrhythmia, myocardial infarction, and death.
   D. **Factors to consider** (*Investigative Radiology* 1S:S32–S39, 1980) are: (1) Severe reactions and death are most common in those above the age of 60 years. (2) Cardiac disease increases the risk 4–5 times. (3) Asthma increases the risk 5 times. (4) History of allergy increases the risk 4 times. (5) Previous reactions increase the risk 11 times. Thus, the most dangerous patient is fearful, obese, older than 50, and a smoker with a history of cardiac disease, asthma, and a claim to be allergic to everything.

II. **Dosage.** Deaths have occured with as little as 0.1 mL injected intradermally and with 5 mL or less. Patients have died as a result of overdosage, volumes of 250–300 mL or more of

Table 14.5. Initial Signs of Reaction Leading to Death during Urography with Ionic Contrast Media[a]

Nausea and emesis	20
Respiratory distress	20
Hypotension	13
Cardiac arrhythmia or arrest	10
Clonic seizures	8
Unconsciousness	8
Chills and fever	7
Apprehension and restlessness	7
Respiratory arrest	7
Severe erythema	4
"Peculiar feeling"	4
Sneezing	3
Chest pain	3
Other various signs (pain, itching, headache)	16

[a]Data are from *Radiology* 134:1–12, 1980.

undiluted ionic media. Test dosing in any form is useless and should not be employed. Use of an adequate but not a high dose is recommended. Contrast media is not as safe as normal saline even in nonionic form.

III. **Causes of death** as determined in 130 necropsies associated with intravenous and intra-arterial use of contrast media were (*Australasian Radiology* 28:133–135, 1984):

Cardiac arrest, myocardial infarction, coronary artery disease, and arrhythmia	51
Pulmonary edema	29
Cerebral edema	11
Stroke	8
Consumption coagulopathy	7
Unknown	7
Respiratory arrest	4
Glottic edema	4
Other	9
Total	130

IV. **Possible explanations** of contrast media reactions
   A. **Allergy** (*Investigative Radiology* 15:529–532, 1980)
   B. **Histamine release** (*Journal of Allergy and Clinical Immunology* 63:281–289, 1979)
   C. **Complement activation** (*Investigative Radiology* 15:2–6, 1980)
   D. **Hyperosmolality** (*British Journal of Radiology* 55:1–18, 1982)
   E. **Central nervous system effect** (*Radiology* 134:1–12, 1980)

V. **Discussion of possible explanations**
   A. **Allergy** is commonly held to be **due to** occurrence of **urticaria** and blamed on iodide content of contrast media. The iodide is firmly bound, and no antibodies to contrast media have been demonstrated in humans or animals.
   B. **Complement activation** explains rare cases of consumption coagulopathy but not cardiac deaths and pulmonary edema.
   C. **Histamine release** is evanescent, is quickly deactivated, and could only explain some cardiac arrhythmias, hypotension, and urticaria.
   D. **Hyperosmolality.** This characteristic of contrast media could explain episodes of hypotension, pulmonary artery hypertension, and several cardiac arrhythmias but not nausea, pulmonary edema, or urticaria.
   E. **Central nervous system.** A coherent and clinically useful explanation of all contrast media reactions depends on the

effect of these media on and through the central nervous system (CNS) (Fig. 14.1).

1. **Mechanism.** The limbic system, a primitive portion of the brain, has intimate and special connections to the hypothalamus and thus controls flight or fright reactions through the autonomic nervous system. Anxiety will enhance this potential response. The feeling of warmth, pain, or nausea may trigger a severe response in a susceptible individual, especially one with severe coronary artery disease.

    a. **Hypotension or shock.** Direct connections from the hypothalamus to the vasomotor center distributed through the autonomic nervous system to the cardiovascular system produce snycope and shock.

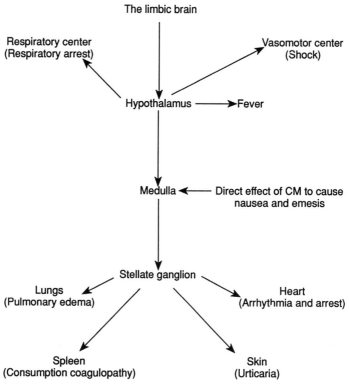

Figure 14.1. Contrast media reactions and the CNS. *CM,* contrast media.

b. **Nausea and emesis.** The known defect in the blood-brain barrier (BBB) in the area postrema, the site of the trigger zone for the nausea and and vomiting center, allows the circulating contrast media to have a direct effect that varies according to the toxicity of the media and the size of the BBB defect.

c. **Fever.** Connections between the anterior and posterior portions of the hypothalamus control body temperature.

d. **Pulmonary edema.** The sympathetic side of the autonomic nervous system may be activated by head injury or an autonomic avalanche to cause contraction of the precapillary and postcapillary sphincters to trap blood in the lungs, causing the pulmonary edema (*Journal of Neurosurgery* 28:112–117, 1968).

e. **Bronchospasm.** A small degree of bronchospasm occurs with all contrast media use as a result of the effect of the parasympathetic portion of the autonomic system affecting the bronchi. This may be severe in asthmatics who have acquired this response to stress (*Radiology* 124:17–21, 1977).

f. **Cardia arrhythmia and arrest.** Activation of the parasympathetic system acting through the hypothalamus may cause bradycardia or arrest. Activation of the sympathetic system will cause fibrillation and myocardial infarction.

g. **Consumption coagulopathy.** Anxiety and hostility shorten clotting time and increase blood viscosity. Sympathetic innervation of the spleen causes release of factor VIII, causing consumption coagulopathy and contributing to myocardial infarction.

h. **Urticaria.** Sympathetic fibers to the skin when activated cause the release of histamine locally, producing both urticaria and erythema.

2. **Evidence for the CNS hypothesis**
   a. LD_{50} of intrathecal contrast media is 0.0014 the intravenous LD_{50} (*Annals of the New York Academy of Science* 78:727–739, 1959).
   b. ED_{50} nonionic contrast media is much higher than that of ionic media (*Acta Radiologica Supplementum* 362: 77–81, 1980).
   c. LD_{50} of mice protected by diazepam is much higher

with intravenous injections of ionic media (*Radiology* 138:47–49, 1981).
   d. Conscious dogs made anxious by suddenly induced complete paralysis have a greater reactivity to intravenous ionic contrast media (*Radiology* 76:88–95, 1961).
VI. **Prevention of reactions**
   A. **Choose patients.** Be wary of extremely fearful patients, especially those older than 50 years with cardiac disease who cannot be won over to your confidence by your interest in them and their problem. Treat all the patients with gentleness, calmness, warmth, and understanding.
   B. **Employ diazepam** or another similar drug that has a specific effect on the limbic brain, in a dose up to 10 mg delivered IV slowly while you are talking to the patient pleasantly (*Radiology* 138:47–49, 1981).
   C. **Use nonionic media,** as it is less toxic and irritating than ionic media and can be used for myelography.
   D. **Corticosteroid protection?** Corticosteroids are commonly used for prophylaxis, but the reasons are nebulous and the beneficial effect is doubtful. There may be a placebo effect, however, since the patient is aware that you are trying to protect him. Deaths and severe reactions have happened despite corticosteroid prophylaxis. A recent study supporting this use is of doubtful validity because of the marked difference in adverse reactions in two placebo groups (*New England Journal of Medicine* 317:845–849, 1987).
   E. **Antihistamines and histamine blockers.** These are of no proven benefit, but IV antihistamines may cause a soporific effect and thus calm some patients.
VII. **Treatment of reactions**
   A. **Maintain the airway.**
   B. **Infuse** large volumes of **IV fluids** to treat shock.
   C. **Give atropine,** at least 0.4 mg **IV,** if bradycardia occurs.
   D. **Give dopamine IV** if tachycardia occurs.
   E. **Use cardioversion** if ventricular fibrillation occurs.
   F. **Give diazepam** if convulsions occur and, on occasion, if the patient is not in extremis.
   G. **Antihistamines and corticosteroids** are useless as initial therapy. Any histamine released is very rapidly deactivated naturally, and an hour or more must elapse before IV corticosteroids can hope to have any effect.

Part III

I. **General considerations.** This chapter is concerned specifically with the systemic reactions that may sometimes accompany the intravascular injections of contrast media. These reactions are diversely labeled as "idiosyncratic," "anaphylactoid," "hypersensitivity," and "allergic." Reactions may occur with all known forms of intravascular contrast media but are more frequent with ionic hyperosmolar media than with nonionic low osmolar media. For example, in the largest comparative study the overall incidence of reactions in patients receiving ionic media was 12.6% and, for those receiving nonionic media, 3.13%. Severe reactions were defined in this study as one or any combination of dyspnea, sudden drop of blood pressure, cardiac arrest, or loss of consciousness and occurred at the rate of 0.22% in the group receiving the ionic media and of 0.04% in the group receiving nonionic media (*Radiology* 175:621–628, 1990). The true mortality rate for contrast media is not known. The most widely quoted rates vary somewhere between 1:40,000 and 1:120,000.

In general, contrast material reactions can be categorized as (*a*) chemotoxic reactions and (*b*) systemic reactions. Chemotoxic reactions are those pathophysiologic abnormalities that may be produced by injections of high concentrations of contrast media into a vessel leading to a vital organ, such as the brain, heart, or kidney. Systemic reactions are principally dose-independent and produce signs and symptoms that mimic true anaphylaxis. From a purely pharmacologic point of view, contrast media should be regarded as among the safest of all intravascular drugs. Nevertheless, the high utilization rate of these substances must invariably be associated with significant numbers of reactions even at a very low reaction incidence.

II. **Chemical composition and toxicity.** The composition of these molecules has been addressed in an earlier chapter. From multiple studies it seems likely that both osmolality and the chemical composition play a role in toxicity considerations. The LD_{50} animal model is far and away the most commonly cited index of contrast material toxicity. Appropriately viewed, however, it is clear that this index correlates more closely with chemotoxicity than with considerations of anaphylactoid systemic toxicity. In fact, there is neither an animal nor an in vitro model developed thus far that correlates satisfactorily with systemic toxicity. An in vitro correlate of the chemotoxicity and LD_{50} level, however, appears to be the binding of specific contrast material molecules to serum proteins. Increased binding correlates with increased toxicity, increased osmolality, and an increased ratio of hydrophobicity to hydrophilicity of individual contrast media (*AJR* 87:338–360, 1962).

The following signs and symptoms are commonly reported in patients experiencing contrast reactions: nausea, vomiting, vertigo, heat sensation, and sneezing; urticaria, alterations in blood pressure, bronchospasm, laryngospasm, and angioedema; and loss of consciousness, convulsions, pulmonary edema, and angina. While this grouping of signs and symptoms might be categorized respectively as mild, moderate, and severe, such categorization in the reporting of reactions tends mostly to obscure potential comparisons between reported studies, since there is no universal agreement on this categorization.

III. **High risk versus low risk.** Since there is currently no laboratory or clinical test that correlates significantly with contrast material toxicity, historical factors constitute the singlemost reliable index of a patient's potential response to contrast challenge. Therefore, a GOOD HISTORY must be obtained PRIOR to any contrast material injection. The best single historical index is a history of a previous contrast material reaction (increases the likelihood of a current reaction 3–5 times). The next best index is a history of extrinsic or intrinsic asthma or a history of previous anaphylactic reactions to any drug, food, or toxin. Next would be patients with any form of antigen-modulated hypersensitivity. Additional patients at increased risk would include patients with incipient cardiac failure, patients with general debilitation, patients poorly hydrated, patients with sickle cell disease, and patients with renal insufficiency who are unable to excrete the contrast material rapidly.

IV. **Pathogenesis of systemic reactions**
   A. **Mediators and activation systems.** Despite considerable study, the exact pathogenesis of systemic reactions is still incompletely understood. Attention has been given to the following potential mediators: histamine, bradykinin, leukotrienes, vasoactive prostaglandins, C3a, and C5a.

   Contrast media have been reported to liberate or activate all of these with the exception of the leukotrienes. In addition, contrast media have been noted to activate the contact, coagulation, and fibrinolysin systems. In early studies, histamine release was considered to be a predominant mechanism. Contrast media can release histamine from mast cells via a nonimmune mechanism (*Investigative Radiology* 5:503–509, 1970). No contrast correlation between histamine levels and symptoms, however, has been established, and it seems unlikely that histamine release per se accounts for the major pathophysiologic alterations seen in patients suffering severe systemic reactions.

B. **Immunoglobulin E (IgE) and contrast media reactions.** A potential role for IgE (immediate anaphylaxis) has been considered in a number of reports. However, contrast media are highly nonreactive chemically, do not demonstrate covalent binding to proteins, and would therefore not be likely candidates to form haptens. A history of a previous contrast reaction does *not guarantee* a current reaction. The great majority of previous reactors *will not* have a current reaction. A previous exposure to contrast material is not necessary for a current reaction. To date, there are no reports of IgE antibodies produced in either rabbits or humans by injections of intravascular contrast media that have not been artificially conjugated with protein.

C. **Is there, nevertheless, a role for IgE in contrast material anaphylaxis?** Patients with various allergies and asthma are at higher risk. While the precise rationale for the higher risk necessitates further study, it is highly unlikely that it is due to a proclivity of these patients to form antibodies to contrast media. It is more likely that nonrelevant antigen processing in these patients leads to the production of mediators that crossconnect with contrast media-relevant activation systems. Thus, nonrelevant (noncontrast material) antigen processing may act as a bolstering reaction in hypersensitive individuals undergoing more fundamental mechanisms of contrast media reactions.

D. **Effects of contrast media on activation systems**
   1. **Complement system activation.** C3a and C5a are physiologically active peptides (anaphylatoxins). These peptides can bring about a wide variety of reactions in vivo (increase capillary permeability, release of histamine from mast cells or basophils, directed attraction of white blood cells, and release of hydrolysis from white blood cells). In vitro incubation of human plasma with contrast

media will result in activation of the complement system (*Investigative Radiology* 11:303–308, 1976; 15:52–55, 1980). In vivo, activation has also been reported. Since high concentrations of contrast media are necessary to produce activation in vitro, it is unlikely that direct exposure of contrast media produces any significant complement activation in vivo. Nevertheless, unequivocal complement activation may occur in patients having reactions (*Investigative Radiology* 15:S6–S12, 1980). In all likelihood, however, this is a secondary phenomenon consequent to a more direct activation of other mediator systems.

E. Plasma contact system. The plasma contact system is a protein cascade that involves sequentially the activation of factor XII, prekallikrein, and high-molecular-weight kininogen. The activation of high-molecular-weight kininogen results in the release of bradykinin. Bradykinin is capable of producing all of the pathophysiologic effects attributed to histamine but is considerably more potent on a mole per mole basis. Activation of this protein cascade occurs when either a polyanionic surface comes in contact with factor XII in plasma or proteases capable of activating factor XII are present. Surfaces capable of activating factor XII have been described in subendothelial areas in mast cells and basophils and in certain other extravascular tissues. Intravascular contrast material injections have the potential to disrupt the endothelial lining of blood vessels (*Investigative Radiology* 23:S147–S149, 1988). This occurs in greater intensity with the ionics than with the nonionic substances. As noted earlier, contrast media can produce a direct (nonimmunologic) release of mast cell contents. Utilizing a synthetic exogenous substance to simulate endogenous "surfaces," it has been demonstrated that the plasmas of contrast material reactors and the plasmas of asthmatics display an accelerated activation of the contact system under certain specific laboratory conditions. Further studies suggested that these accelerated activations might be due to the endogenous presence of primers capable of potentiating contact system activation (*Thrombosis Research* 48:729–736, 1987).

In common with other activation systems, the contact system has a number of important inhibitors that help to maintain homeostasis. The most important of these is a substance that also has major responsibility for inhibition of the initiating protein in the complement cascade (the C1-esterase inhibitor). Increase in the concentration of this inhibitor has been described after corticosteroid administration (*Investigative Radiology* 16:20–23, 1981). This may be one factor that plays a role in the diminished incidence of reactions occurring after corticosteroid pretreatment (see subsequent sections).

Bradykinin, the end product of the contact system activation, has a limited half-life in the circulation because of the presence of two enzymes capable of hydrolyzing bradykinin. These are, respectively, angiotensin-converting enzyme (kininase II) and carboxypeptidase N (kininase I). It is of theoretical interest that studies of kininase II have shown that the contrast media, in sufficient concentration, have the potential to inhibit this kininase partially.

The contrast material-induced activation of factor XII, presumed to result from endothelial disruption, may also produce activation of the intrinsic coagulation system. In some individuals, this (rarely) may lead to a disseminated intravascular coagulation.

Finally, it should be noted that bradykinin, in addition to its histamine-like effects, may produce secondary effects that are even more profound. These result from the bradykinin-induced activation of arachidonic acid in some cellular membranes, fueling both the cyclooxygenase and lipooxygenase pathways and resulting in the potential release of vasoactive prostaglandins and leukotrienes.

V. **Role of anxiety.** There is some belief that central nervous system events, particularly anxiety, play an important role in contrast material anaphylaxis (*Radiology* 138:47–49, 1981). It is certainly possible that anxiety may promote the release of some circulating mediators, and there is no doubt that classic vagovaso or vagovagal mechanisms may play a role in reactions. However, informed consent does not appear to alter the incidence of reactions, and there is certainly no experimental support for the notion that all contrast material reactions are related to activation of the limbic portion of the central nervous system.

VI. **Patients at risk: the question of pretreatment.** At the present time, there is no reliable in vitro test or contrast material pretest that will reliably project the likelihood of a contrast-induced reaction. Therefore, a careful **physician-obtained** history is imperative. There is a definite increased risk in patients who have a history of a previous contrast reaction and/or a history of significant allergy or asthma. Patients with a history of a sensation of heat, flushing, or a single episode of nausea or vomiting need not be considered as "having had a previous reaction."

Other patients may be at risk, not because there is an increased likelihood of having a reaction, but because the occurrence of a reaction at its usual incidence might be particularly harmful. Such patients include those with generalized severe debilitation, those with cardiac dysfunction including recent or potentially imminent decompensation, and those with sickle cell disease.

In patients thought to be at increased risk of a reaction and in patients thought to be at particular risk because of a reaction, pretreatment may be considered. Pretreatment, when administered, has consisted generally of corticosteroids alone, antihistamines alone, or a combination of these two. In addition, beta-adrenergic agonists have been utilized. Only one blinded and randomized study of the effectiveness of pretreatment has been reported (*New England Journal of Medicine* 317:845–849, 1987). In this study, it was determined that the utilization of oral corticosteroids, given 12 hours and again 2 hours prior to contrast challenge, effectively diminished the incidence of mild, moderate, and severe reactions. Corticosteroids given in this dosage appeared to be indistinguishable from placebo insofar as side effects from the corticosteroids alone were considered. A 32-mg tablet of Medrol (methylprednisolone) was utilized in these studies. It was also determined that a single oral 32-mg dose of Medrol given just 2 hours prior to contrast challenge was indistinguishable from placebo in terms of protection. Corticosteroids, to be effective, must be given some hours prior to contrast challenge. Unfortunately, the minimum lead time is not known but is unlikely to be less than 6 hours. This is in keeping with the pharmacology of this drug that involves complexing with a cytoplasmic receptor, migration of the complex to the cell nucleus, transcription of the complex "message" to activate a DNA-dependent synthesis of RNA, and subsequently, the accelerated formation of specific enzymes, inhibitors, etc. These reactions take **time**. Since both complement activation and contact activation have been implicated in contrast reactions, the corticosteroid induction of the major inhibitor of these two cascades (the C1-esterase inhibitor) may be significant. The production of vasoactive prostaglandins and the production of leukotrienes are dependent on the mobilization of arachidonic acid from cell membranes. Bradykinin, the end product of contact activation, stimulates such mobilization. Corticosteroids have the opposite effect. These and possible other effects, then, may account in part for the protective effects of corticosteroid pretreatment. Histamine release, although it may play a minor role in reactions, does occur as a consequence of contrast material injections (*Radiology* 110:49–59, 1974). Thus, there is a rationale for the utilization of antihistamines as pretreatment. The bulk of experience has been with H_1-blockers. A possible role for H_2-blockers, however, cannot be ruled out as of this time. Likewise, the role of beta-agonists is not yet definitive. Nevertheless, several published studies from one center indicate that the addition of ephedrine sulfate, in a 24-mg oral dose, 1 hour prior to the procedure, conferred additional protection when added to a regimen of prednisone and antihistamines (*Journal of Allergy and Clinical Immunology* 74:540–543, 1984).

VI. **Treatment of reactions: treatment based on consideration of relevant pathogenic concepts.** Six classes of drugs have been utilized in the treatment of reactions. Prior to the administration of these drugs and/or concurrent with them, consideration should be given to the utilization of intravenous fluids and, if appropriate, nasal oxygen. The rationale for the utilization of these is self-evident, since a fall in blood pressure or bronchospasm is often the most obvious event in moderate-to-severe reactions. All of the mediators mentioned in the preceding sections are capable of producing these two effects. If the reaction is sufficiently severe and/or does not respond to intravenous fluids and nasal oxygen, consideration can be given to the utilization of drug therapy. The major classes of drugs utilized in these reactions include: (1) adrenergic agonists, (2) bronchodilators, (3) anticholinergics, (4) antihistamines, (5) steroids, and (6) anticonvulsants. There are no controlled studies of the treatment of severe anaphylactic or anaphylactoid reactions resulting from the administration of contrast media or exposure to other drugs or antigens. Nevertheless, epinephrine is generally considered to be the first line of defense. If the reaction is mild to moderate, epinephrine can be given subcutaneously in dosages of 0.1–0.2 mL of a 1:1000 mixture. For more severe reactions, epinephrine should be given intravenously. When administered via this route, the injection should be given slowly, and the epinephrine should be given in low concentration (approximately 0.1 mL of 1:10,000 solution per minute). If administered by drip, a rate of about 10 µg/min should be utilized, and the rate of injection titrated with clinical response. The utilization of this drug assumes protective actions of both the alpha-agonist and beta-agonist components. The alpha-agonist effects of epinephrine increase blood pressure and reverse peripheral vasodilatation. These vasoconstrictor changes may also decrease angioedema and urticaria. The beta-agonist actions of the drug reverse bronchoconstriction, produce positive inotropic and chronotropic cardiac effects, and may increase intracellular cyclic AMP. Increments in cyclic AMP levels are generally associated with an inhibition of mediator release. The following caveats should be kept in mind. Beta-receptor sites ordinarily respond at lower doses of epinephrine than alpha-receptor sites, but a patient on beta-blockers may show a refractory response at the beta-receptor site, encouraging the utilization of an epinephrine dose to the point that individuals with a fragile intracerebral or coronary circulation may experience a hypertensive crisis, resulting in a stroke or myocardial ischemia. Remember also that patients with asthma or certain other significant allergic diseases may demonstrate an endogenous hypo-beta-adrenergic response.

If beta-adrenergic responsiveness, either drug- or disease-induced, is sufficiently low, there may also be a dominance of cholinergic activity and, possibly, a secondary bradycardia. In these circumstances, consideration can be given to the utilization of isoproterenol, a beta-agonist, and/or to the utilization of anticholinergics. The effective use of atropine in a number of patients suffering contrast material anaphylaxis with bradycardia has been reported (*Radiology* 121:5–7, 1976).

Beta-agonists can also be given by metered dose inhalers (e.g., Alupent) for isolated bronchospasm or particularly for bronchospasm concomitant with other symptoms of anaphylaxis.

Antihistamines are, in fact, the most commonly administered drug for the treatment of contrast reactions. Nevertheless, the rationale for the utilization of these drugs for the treatment of reactions is questionable, since, presumably, the salient histamine receptors have already been occupied. Here again, one cannot fall back on controlled studies, and while the utilization of H_1-blockers has been widespread, the utilization of H_2-blockers must rest on a much smaller, uncontrolled experience.

If a patient develops protracted seizures, the administration of anticonvulsant drugs must be considered.

The utilization of steroids in the treatment of severe reactions has been advocated. Here again, there are no hard data. From previous discussion, it is clear that no immediate effect should be expected, but possible protection over time might be significant, since some (very few) severe reactions will extend over a matter of hours rather than minutes. If corticosteroids are to be utilized as treatment, they should be administered in large intravenous dosages.

Chapter / 15
Low Osmolar Contrast Agents: Economic and Legal Issues

C. John Rosenquist
Peter D. Jacobson

I. Introduction
A. The proposed development of low osmolar contrast media (LOCM) by Torsten Almén in 1968 was a major advance in providing safer and better tolerated agents for radiographic examinations (*Journal of Theoretical Biology* 24:216–226, 1969). Since that time, evidence has accumulated that documents the increased safety of LOCM. There is now general agreement that LOCM would be used for all contrast studies if there were no financial concerns. Because LOCM cost 3–12 times more than high osmolar contrast media (HOCM), debate has continued about the appropriate utilization of the low osmolar agents. The issue of cost effectiveness of LOCM is discussed in the first section of this chapter.
B. Because of the increased cost of LOCM compared with HOCM and because of resulting questions about appropriate use, several interesting legal issues have developed. First is the question of negligence. If LOCM are safer than HOCM, does the continued use of HOCM constitute medical negligence? Second, should the patient be involved in the decision about what contrast agent to use? Does informed consent include informing the patient about a choice of contrast media and the associated benefits and risks? These issues are analyzed in the second section of the chapter.

II. Economic issues
A. **Analysis of cost.** Although there is considerable variation in cost of LOCM in different countries, the increase in cost of LOCM compared with HOCM has become a global issue. Currently, the cost ratio for LOCM to HOCM ranges from 3:1 in most European countries to approximately 13:1 in the United States. Why there is this difference in cost is not

entirely clear, but reasons given for the greater cost in the United States include expenses incurred to gain Food and Drug Administration (FDA) approval, FDA-required clinical tests, royalties paid to companies that initially developed the agents, and increased manufacturing costs. There is little evidence thus far that introduction of additional manufacturers in the market will decrease costs or that increased use will result in economies of scale.

The conversion to total use of LOCM would add a significant burden to the cost of medical care. In the United States alone the increase in cost would be about $1 billion. This must be evaluated in the context of the total annual health care cost of $600 billion. On a global scale, it has been estimated that the increase in cost for universal use of LOCM would be about $4 billion.

During the past several years the percentage of patients receiving LOCM has been increasing in all countries. However, that percentage varies greatly for several reasons. First, as one might expect, usage is greater in those countries in which the price differential is lower. Second, decisions about coverage by third-party payers have led to differences in use. Finally, a concern about possible legal complications if LOCM are not used has led to greater use in some locations. It is estimated that currently about 25% of patients receive LOCM in the United States, while in Japan and West Germany about 70–80% of patients receive LOCM.

One additional cost factor must be considered. Since the incidence of minor and major reactions is decreased with LOCM, there is a cost saving in treatment of these reactions. Minor reactions require little if any treatment and do not require hospitalization. Although major reactions do add to the cost of an examination, they occur in only 0.1–0.2% of injections in low-risk patients, and the added cost does not offset the cost of LOCM for every examination (*Radiology* 169:163–168, 1988).

B. **Analysis of benefit.** Because of the low incidence of reactions with both LOCM and HOCM, studies of large numbers of patients are needed to prove that the newer agents are safer. This is especially true of fatal reactions, which are estimated to occur in approximately 1 in 40,000 for HOCM injections, to 1 in 250,000 for LOCM injections. There have not been any randomized studies that have documented this difference in reaction rates.

In the past 2 years, two studies involving large numbers of patients have demonstrated a significant difference in rates of minor and major reactions between HOCM and LOCM (*Australasian Radiology* 32:426–428, 1988; *Radiology* 175: 621–628, 1990). Both studies revealed a 3–5-times decrease in reactions with LOCM. Neither study demonstrated a difference in the incidence of fatal reactions.

An additional well-proven benefit of LOCM is the decrease in pain associated with intra-arterial injection. This advantage is difficult to quantitate as part of an analysis of the usefulness of the new agents. One study attempted to do this by asking patients how much they would be willing to pay in order to avoid the pain and increased risk of reaction associated with HOCM. Of course, this is difficult for a patient to assess prior to the injection.

C. **Cost-effectiveness analysis.** There is now universal agreement that LOCM should be used for some percentage of contrast studies. The issue to be resolved is what group should receive these new agents or whether all studies should be performed with LOCM. By combining the information discussed in the sections about costs and benefits, it is possible to analyze the use of LOCM and compare the cost effectiveness of use of these media with that of other accepted medical treatments.

It has been estimated that if LOCM were used for all contrast studies in the United States, the additional cost would be $1 billion/year (*JAMA* 260:1586–1592, 1988). It has been suggested that LOCM be used only for the group of patients having an increased risk of reaction to contrast media. Patients usually included in this group include those with a history of allergies or previous reaction to contrast media, diabetes, renal or cardiac disease, and infants and those above age 65 years. Depending on the specific criteria used and the patient population, the high-risk group may be 20–40% of all patients. If the use of LOCM is limited to this high-risk group, the annual increase in cost in the United States would be $176 million, and the cost per year life saved would be $31 thousand. If LOCM were used for all injections *rather than* the high-risk group, the additional or marginal cost would be about $850 million/year, and the additional cost per year life saved would be $234 thousand.

This type of cost-effectiveness analysis is useful for determining how best to spend the limited resources that are available for medical care. By limiting the use of LOCM to high-risk patients, the cost per year life saved ($31,000) is similar to other medical programs that now exist. **The issue is not whether the United States or other countries can afford to provide LOCM for all contrast studies but rather how best to utilize the funds that are available.** Because of the rapidly increasing cost of medical care, there is general agreement that such decisions will become more important, and cost-effectiveness analysis provides a rational basis for developing appropriate health policy.

II. Legal considerations

A. **Negligence.** In the absence of established legal precedent or professional guidance, hospitals and physicians are confronted with a difficult choice in determining the appropriate use of LOCM: the need to balance the high costs of universal LOCM use with potential legal liability for improperly limiting its use. Under what circumstances will failure to use LOCM result in a medical malpractice award? More specifically, when does a failure to use a new, safer technology (LOCM) constitute a deviation from customary medical practice?

To prove liability for medical malpractice, the injured patient must first show that the physician failed to exercise the appropriate standard of care owed to that patient. In medical malpractice cases, the medical profession sets its own standard of care based on what is customary and usual medical practice, as established through medical testimony and medical treatises. Courts are reluctant to substitute their judgment for that of the medical profession, even when a new, safer technology is being considered.

Each physician must exercise the degree of skill ordinarily practiced, under similar circumstances, by members of the profession. Physicians with special knowledge, such as radiologists, will be held to customary practices among those of equivalent skill and training. If, however, there is more than one recognized course of treatment, most courts allow some flexibility in what is regarded as customary (the respectable minority rule). The injured patient must also prove that the failure to maintain the standard of care caused the injury and that damages were incurred.

One source of potential hospital liability for new technology is the facility's duty to provide adequate equipment and medical services. In general, hospitals are not required to maintain the latest technology, as long as its equipment is safe, is generally used by other hospitals, and functions as designed. **The availability of a new technology alone, even if safer, does not prove that the existing technology is ineffective or unsafe.**

If use of LOCM becomes the customary practice for all contrast injections, including those for both high- and low-risk patients, a severe or fatal reaction following use of HOCM could result in a medical malpractice award. Because, however, the use of LOCM is currently not the standard of care for all contrast injections, different possibilities must be considered. At this point, LOCM are being increasingly used for identified high-risk patients, while HOCM remain customary for low-risk patients. Thus, liability concerns should be separated based on the patient's risk status.

As the evidence of the safety and efficacy of LOCM for high-risk patients mounts and as LOCM are used increasingly for them, the failure to use LOCM for high-risk patients is likely to be considered negligent. In view of the recent American College of Radiology statement that LOCM should be considered for defined high-risk patients (such as those with diabetes, heart problems, or a prior adverse contrast reaction), it may be difficult to justify using HOCM for high-risk patients. At the very least, the physician would have the burden of justifying the failure to use LOCM.

A much more difficult situation is presented when a low-risk patient suffers a severe adverse reaction from an HOCM injection. Assuming, first, that customary practice is to use HOCM for low-risk patients, is it nevertheless negligent not to use LOCM? In limited circumstances, courts have rejected customary practice and determined that the failure to use a safer technology is indeed negligent, but the availability of LOCM presents a very different situation from those cases. The magnitude of projected expenditures for the new media relative to projected benefits, particularly for low-risk patients, suggests that courts would be reluctant to second-guess the medical profession. As long as the existing contrast media are safe and not defective, it seems unlikely that the availability of the new contrast media alone would be sufficient to shift the customary practice standard.

Let's suppose, however, that the medical profession, through clinical practice guidelines, recommends that LOCM should be limited to high-risk patients. Would this increase the probability that a low-risk patient could successfully bring a malpractice lawsuit for failure to use LOCM? The question is whether a court should disregard the limits set by the profession in defining what is customary care. Again, it is unlikely that courts would impose liability, given the substantial cost disparities between LOCM and HOCM and the limited incremental safety benefits for low-risk patients. Because nothing prevents the profession from factoring in resource constraints in defining the level of technology that will become customary practice, this would be strong evidence of customary practice that courts should uphold. Although it is possible that in reaching a decision courts might disregard the notion of aggregate costs and focus instead on the small episodic cost of LOCM, it is still likely that courts would be reluctant to impose liability. Even if episodic costs are focused on, the cost of LOCM will probably exceed the expected loss (the probability of harm times the expected award) for low-risk patients, hence the failure to use LOCM would not be negligent.

B. Informed consent. The issue of informed consent is an important consideration for radiologists in contemplating the introduction of new contrast media. In general, the law of informed consent has evolved to create an independent basis for liability over the failure to inform patients properly about the risks of certain treatment methods and the availability of suitable alternatives so that the patient can make an informed decision to undergo or reject the procedure. Just what constitutes adequate risk disclosure is, of course, the operative question and can only be decided on a case-by-case basis. At a minimum, physicians must inform patients of all "material" risks, which courts apply by emphasizing the severity of the risk and its incidence, and available treatment alternatives.

There are basically two standards for determining the specific rules regarding informed consent. In states adopting the reasonable physician standard, physicians are required to inform patients of material risks and alternatives, based on what a reasonable physician would expect. In states adopting the reasonable patient standard, informed consent requirements are based on what a reasonable patient would expect. Injured patients have an easier time winning a lawsuit in those states using the reasonable patient standard.

Given that LOCM are available for high-risk patients, is a physician required to inform low-risk patients of its availability? This depends on the answers to two questions: If an alternative is limited to high-risk patients, can a physician be held liable for failure to inform a low-risk patient of the alternative? And, is the availability of LOCM, regardless of cost or who pays, an alternative that must be considered? In other words, should a patient be informed of more costly alternatives, even those that are not reimbursable or that the medical profession limits because of cost?

No court has considered whether alternatives to be disclosed include procedures limited by cost or by customary practice. Clearly, the patient must be informed of all anticipated material risks and other reasonably available therapeutic alternatives. But it is doubtful that every potential alternative, regardless of its proven safety and effectiveness or cost should be evaluated by the patient. For example, patients generally need not be informed of futile or useless alternatives. If a procedure is not customary practice, it should not be considered as a viable alternative. Alternatives that are prohibitively expensive or restricted to certain high-risk categories should not be part of informed consent. Without much doubt, requiring informed consent for LOCM is tantamount to distribution for all contrast injections because third-party insurance would cover most of the costs. It should be clear, however, that some scholars would require a more expansive view of informed consent requirements (*AJR* 151:529–531, 1988), and the American College of Radiology has suggested that LOCM should be used if a patient requests it.

In summary, the failure to use LOCM for low-risk patients should not result in liability awards unless such use becomes customary practice. Although some physicians and hospital administrators may have switched to universal LOCM use because of liability concerns, there is no evidence that these concerns are currently justified. **As long as customary care is to use LOCM for high-risk patients and HOCM for low-risk patients, it should be legally acceptable to use HOCM for low-risk patients.**

Chapter / 16
Barium Sulfate for Gastrointestinal Use

Jovitas Skucas

I. Clinical application
A. Basic properties.
Barium sulfate is a white crystalline powder having a molecular weight of 233. It is essentially insoluble in water. Its specific gravity of 4.5 results in the not uncommon patient observation that a cup of barium suspension is "heavy."

Barium sulfate is found in deposits scattered throughout the earth. Although these deposits are extensively mined, the presence of toxic impurities generally limits the use of such mined barium sulfate. Barium sulfate designed specifically for radiographic use is generally obtained by a number of precipitation processes from other compounds.

Depending on the precipitation process, the barium sulfate particles can be made in a variety of sizes. Particles approximately 0.5 μm in diameter are generally used as additives in other formulations. The larger sizes, from 5 to 12 μm in diameter, are used in commercial "high-density" products. In fact, some of the barium products specifically designed for double-contrast gastric studies can contain a significant number of particles 18 μm or larger in diameter. Generally, the products designed for gastric coating have an extreme heterogeneity in particle size, while products designed for single-contrast studies tend to be more homogeneous.

Any suspension of barium sulfate will eventually settle out. Larger particles settle faster and form a more "dense" cake than smaller particles. It is thus not unusual that the larger particle double-contrast barium products tend to form a relatively hard cake at the bottom of a storage jug and considerable shaking is required to force the particles back into suspension. Because sedimentation also occurs once the suspension is poured into individual patient cups, these should be filled only prior to immediate use. The barium manufacturers try to decrease sedimentation through use of various additives, generally with varying degrees of success.

The large-particle, high-density barium suspensions designed for double-contrast examinations should not be diluted for use in single-contrast studies. Such dilution leads to rapid sedimentation in the gastrointestinal tract, with the result that a portion of the nondependent lumen will contain little barium. The products designed primarily for single-contrast work, on the other hand, can be diluted considerably before any settling occurs, a factor due primarily to the relatively small barium particles being used.

Flocculation is a chemical process that results in a coarse precipitate of barium particles. It is not the same as sedimentation. Flocculation can be decreased by a number of protective agents. Over the years, the art of commercial barium sulfate product preparation has evolved to the point where, today, flocculation is only a minor problem with the commercially available products.

Viscosity of the various commercial products varies considerably. Most barium sulfate products exhibit nonnewtonian flow; the viscosity is thus not identical at different flow rates. The viscosity not only determines the flow rate through tubing but also has a major influence on the subsequent mucosal coating thickness. Ideally, a thick mucosal coating is desired; unfortunately, viscosity can be increased only to a certain point before the product forms a paste and coating properties are degraded.

Many of the commercial preparations contain **preservatives.** Although barium sulfate by itself is inert and does not support bacterial growth, a number of the additives are organic in nature. **Once a container is opened or reconstituted with tap water, the suspension should be refrigerated if it is to be kept overnight.**

It is obvious that an ideal barium suspension equally applicable throughout the gastrointestinal tract has not yet been developed. The suspension should not be too fluid; it should coat the mucosa with a sufficiently thick coating to be adequately visualized on radiographs. Once coated, the barium film should remain adherent to the mucosa for a sufficiently long period of time to allow completion of the study.

The terms "thick" and "thin" apply only to the viscosity of a product. They should not be misused to infer radiodensity, which is the result of many other factors.

Three standardized systems are available in measuring the amount of barium sulfate present per unit volume. These are: **specific gravity, weight to volume,** and **weight to weight.**

When the **weight-to-volume** system is used, a certain weight of barium sulfate is added to sufficient water to obtain a predetermined total volume. For example, a 20% weight-to-volume suspension is prepared by adding 20 gm of barium sulfate to enough water for a total volume of 100 mL.

With the **weight-to-weight** system, a certain amount of barium sulfate weight is added to enough water to obtain a predetermined final weight. As an example, a 20% weight-to-weight suspension is prepared by adding 20 gm of barium sulfate to 80 gm (80 mL) of water; the total weight is 100 gm.

In the United States, generally the weight-to-volume or the weight-to-weight system is used. Throughout Europe, the specific gravity method is generally used. Although all three systems are interrelated, they are not easily interchangeable. The differences are especially marked at the higher densities. Conversion tables are available in the literature (*Radiographic Contrast Agents,* 2nd ed. Gaithersburg, MD: Aspen Publishers, 1989, pp. 14–17).

The commercial barium preparations are used on a volume basis, and therefore, cost should also be compared on a volume basis. The cost per kilogram of dry powder can be misleading. While cost per examination is a convenient index, the cost of associated supplies, such as tubing and straws, should be included in this final price. Thus, although the prepackaged liquid products generally cost more per volume than the corresponding bulk powder, once the cost of mixing and accessories is included, quite often the liquid formulation is comparable or even cheaper. In spite of any cost comparisons, however, the most important factor in determining which product to use is the resultant examination quality.

B. **Effervescent agents.** By far, the cheapest second contrast agent is air. In a number of applications, such as double-contrast barium enemas and upper gastrointestinal studies performed through nasogastric tubes, air is the preferred agent. At times, excellent double-contrast esophageal views can be obtained when the patient swallows air together with the barium preparation. Having the patient drink a barium suspension through a large-bore straw that contains side holes helps draw air in (*Radiology* 119:1–5, 1976).

In the past, a commercial preparation had carbon dioxide added to the barium suspension; once the can was opened, the patient drank the "bubbly barium," resulting in release of carbon dioxide in the esophagus and stomach. The effect was similar to drinking a bottle of club soda. With such a product the amount and the speed of gas production were slower than with a conventional effervescent agent (*British Journal of Radiology* 50:546–550, 1977).

Carbon dioxide has been used for double-contrast barium enemas; carbon dioxide is absorbed faster than air, and it is believed that its use results in greater patient comfort (*Radiology* 162:274–275, 1987). Whether a gas such as carbon dioxide or air is used probably does not influence the overall quality of the examination.

Water can also be used as a second contrast. In general, water tends to "wash off" the barium sulfate particles adhering to the mucosa. Part of this effect can be eliminated by using a 0.5% or greater methylcellulose solution.

A number of effervescent tablets, granules, and powders are commercially available. Their overall effect is to produce carbon dioxide on contact with water, and most are satisfactory in producing adequate gastric and duodenal distension. There is considerable variation in the dissolution time between the various products, and whenever a new product is being evaluated, it is best to experiment with the amount to be used until a satisfactory result is achieved.

Most of the commercial effervescent powders and granules come in single-dose packages. In general, the patient places the effervescent agent in the mouth and uses small amounts of water to wash it down. Double-contrast views of the esophagus are obtained immediately. The swallowed gas is subsequently used to obtain double-contrast stomach and duodenal radiographs.

Liquid effervescent agents are available commercially or can be prepared locally by a hospital pharmacy. One commercial manufacturer has a built-in measuring cup for accurate volume dispension of the separate acid and base solutions. In general, the acid consists of citric or tartaric acid, and the base portion is sodium bicarbonate. A dose of 12–15 mL has been found satisfactory for most patients.

Most commercial preparations lead to excessive gas bubbles that can interfere with subsequent diagnosis. Regardless of what contrast agents are being used, if excessive bubbles are encountered, an antifoaming agent should be added. Although many commercial preparations already include such an agent, in many localities it is not sufficient. A commonly used antifoam agent is dimethyl polysiloxane (simethicone). The addition of 1.5 mL of simethicone (equivalent to 100 mg) is often sufficient.

C. **Upper gastrointestinal tract.** A number of authors advocate a study of the esophagus consisting of single-contrast, double-contrast, mucosal relief, and fluoroscopic evaluation of motility (*Radiology* 147:65–70, 1983). Small lesions can be missed by either a single-contrast or a double-contrast study.

Although in some patients with poor esophageal motility or gastroesophageal reflux, sufficient gas is introduced into the esophagus, in most patients a barium suspension and a negative contrast agent are used to obtain satisfactory double-contrast radiographs. **The high-density, low-viscosity barium preparations developed for the stomach are also best for study of the esophagus.** Normal esophageal tonicity leads to esophageal lumen collapse once the primary bolus has passed. Therefore, regardless of what method is used, the study must be performed with reasonable dispatch.

One method found useful is to have the patient drink in quick succession first one and then the other liquid effervescent solution, followed immediately by 60–120 mL of an appropriate barium suspension. Mixing of the two effervescent agents in the esophagus results in distension, followed by subsequent coating by the barium suspension.

If the sequence of ingestion is reversed, namely, the barium is given first, followed by the effervescent agents, visualization of the esophagus is somewhat impaired. On the other hand, "better" gastric mucosal coating is obtained if the barium suspension is given first. Thus optimal coatings in the esophagus and stomach are achieved by varying the sequence of ingestion. In actual practice, the sequence of ingestion used depends on the patient's primary symptomatology; if disease is suspected in the esophagus, the effervescent agents are given first; with suspected gastroduodenal disease, the barium suspension is given first.

Esophageal varices are best studied with the esophagus collapsed. Varices tend to distend with a decrease in intraluminal pressure. If the esophagus is thus maintained in a collapsed state for some time, the chance of detecting small varices is increased. On the other hand, barium coating of the esophagus decreases with time. Although the high-density, low-viscosity barium products will detect some esophageal varices, use of a **barium paste** is preferable. A number of such commercial pastes are available. Some of these are too viscous and flow in a bolus. Overall, the paste viscosity should be similar to that of honey.

One indication for an esophagram is to detect a foreign body. Occasionally, with the patient upright, the weight of a barium column will dislodge such a foreign body into the stomach. In some patients the addition of liquid effervescent agents results in sufficient pressure and esophageal distension that a foreign body passes into the stomach (*Radiology* 146:299–301, 1983).

Barium sulfate tablets can be useful in evaluating **marginal esophageal strictures.** Commercial tablets having a diameter of 12.5 mL are available. These contain 650 mg of barium sulfate plus a number of additives. The tablets are designed to dissolve in either the esophagus or the stomach. Relatively fresh tablets should be used because with age the rate of dissolution can be prolonged (*Radiology* 122:835–836, 1977).

For a single-contrast stomach examination, barium densities varying from 35% to 80% weight to volume are common. Because of swallowed air, some double-contrast radiographs can be obtained. Various compression paddles are of help in decreasing the amount of barium in the path of the X-ray beam.

For the **double-contrast** portion of the examination, the **high-density** and **low-viscosity barium** preparations specifically designed for the upper gastrointestinal tract produce best results. Densities up to 250% weight by volume are commonly used. A volume of 60–120 mL is common. A good barium formulation should result in routine identification of the area gastrica. Once the double-contrast views are obtained, a **lower density barium** suspension is given for subsequent **single-contrast radiographs.**

D. **Small bowel.** There are four radiographic techniques in studying the small bowel: conventional **antegrade examination, enteroclysis, retrograde small bowel examination,** and **peroral pneumocolon.** The choice of examination depends on the clinical indication. Each examination requires modification of the type of barium suspension used and possible inclusion of a second contrast agent.
  1. **Antegrade small bowel examination** is the simplest and the traditional way of studying the small bowel. The patient ingests a barium suspension, and serial views of the small bowel are subsequently obtained. The primary requirement of the contrast agent is that it does not flocculate or precipitate. A 40–60% weight-by-volume suspension is common. The volume required is controversial, with many radiologists preferring 500–800 mL of the suspension. **The contraindications to such a barium study are suspected colonic obstruction or bowel perforation.** Small bowel obstruction is not a contraindication. Barium proximal to a small bowel obstruction continues to stay in liquid form. **Barium inspissation does not form proximal to a small bowel obstruction.** Thus, even in a setting of a small bowel obstruction, use of a large volume of contrast media is safe.
  2. **Enteroclysis.** A number of steerable tubes are available commercially. These are maneuvered through the pylorus into the duodenum or the proximal jejunum, and the contrast agents are subsequently injected through the tube. The rate of injection can be controlled, and the natural flow limiting function of the pylorus thus bypassed. The barium suspension can be infused by gravity, hand-held syringes, or a variety of infusion pumps. Although the rate of infusion can be varied precisely with an infusion pump, in most practices such an elaborate system is not needed, and injection by syringes should be adequate.

     There still is controversy whether a single-contrast or double-contrast enteroclysis study is preferred. Although a double-contrast study tends to be more aesthetically appealing, it is debatable whether the accuracy of the examination is improved (*ROFO* 143:293–297, 1985). For the double-contrast portion of the examination, most American investigators prefer a 0.5% or greater solution of **methylcellulose** in water. The total volume of the two contrast agents can vary considerably. In some patients, up to 2 L of total volume may be required. Most investigators instill the contrast agents until either a lesion is reached or contrast reaches the terminal ileum. If excessive peristalsis is encountered, glucagon can be given to produce hypotonia.

Air as a second-contrast agent is used by a number of Japanese and European investigators. Air results in considerably more radiographic contrast than that seen with a methylcellulose solution.

Although the liquid methylcellulose solution helps to propel a barium suspension ahead of it, a similar effect is not obtained with air; rather, air tends to percolate through the barium-filled loops of small bowel. Some investigators believe that small fistulas and polyps are not as well seen if air is used (*Gastrointestinal Radiology* 1:355–359, 1977). Air bubbles may also lead to confusing shadows. On the other hand, when there are numerous overlapping loops of bowel, such as in the pelvis, infusion of air can be very useful.

There are commercial barium sulfate preparations specifically designed for enteroclysis. For a single-contrast study, a barium sulfate suspension having a specific gravity of approximately 1.27 (equivalent to 34% weight to volume) is preferred. Typical infusion rates are 75–100 mL/min, but the flow rate should be individualized. If the rate is too slow, excessive peristalsis will be present, and the study will be similar to that of a conventional small bowel examination. If the flow rate is too fast, overdistension results in small bowel atonia.

For a double-contrast enteroclysis study, a barium suspension of a higher specific gravity is preferred. A range of 50–95% weight to volume is typical.

3. **Retrograde small bowel examination.** As the name implies, a single-contrast barium enema is performed, and the examination is continued retrograde into the small bowel. Because flow can be controlled by the examiner, the ileum can be readily studied without overlapping loops from more proximal small bowel. A 20% weight-to-volume or somewhat greater concentration is typical. Premedication with a hypotonic agent such as glucagon helps provide patient comfort and also tends to relax the

ileocecal valve. The barium suspension is instilled until the area in question is reached. If a redundant sigmoid colon overlaps the small bowel loops in question, a saline solution can then be infused; such a solution pushes the barium more proximally and achieves a "see-through" effect through the sigmoid colon.

It is common to have a double-contrast study of the terminal ileum whenever a double-contrast barium enema is performed. Such a study is very useful in suspected distal ileal Crohn's disease or in gynecologic malignancies involving the ileum.

4. **Peroral pneumocolon.** This study consists of both antegrade and retrograde components. Initially, a conventional antegrade small bowel examination is performed. Once barium outlines the terminal ileum, air is instilled through the rectum to obtain a double-contrast examination of the distal small bowel. This study can also be combined with enteroclysis.

E. **Colon.** The large bowel can be studied with either a **single-contrast** or **double-contrast** technique. Some radiologists perform a single-contrast study in elderly and debilitated patients. With most double-contrast techniques the patient must lie prone at least part of the time; thus if the patient cannot lie prone, a single-contrast examination is performed.

Disposable prefilled enema bags are commercially available. Some contain the barium in powder form, and the correct amount of water simply is added prior to the examination. Most of these bags must be shaken vigorously to achieve adequate wetting of the barium particles. After such shaking, the bags should be stored on their side; otherwise, considerable settling can occur.

1. **Single-contrast barium enema.** A 12–25% weight-to-volume barium suspension is commonly used for a single-contrast examination. The primary barium suspension requirement is that it not flocculate or settle during the time that the examination is in progress. Because the tendency to settle depends, in part, on the additives present, some products that are well suspended at higher densities settle readily when diluted. If there is any doubt about a particular product's settling rate, a radiograph obtained with a horizontal X-ray beam will show the

degree of settling. Most commercial products designed for single-contrast study consist of rather small barium sulfate particles.

2. **Double-contrast barium enema.** The barium sulfate suspension designed for double-contrast study should be **relatively dense** but still have a sufficiently **low viscosity** so it flows readily through enema tubing. The product should result in uniform mucosal coating without artefacts. The product should not dry out while the examination is being performed.

The barium products designed for double-contrast barium enemas are generally 60–120% weight to volume. Of necessity, their viscosity is also greater than with the lower density barium products designed for single-contrast studies.

Both dry and liquid commercial products are available. If the dry barium powder is used, the amount of water added and the degree of subsequent shaking to achieve wettability should be standardized. In particular, the level marking generally inscribed on the bag should not be used to gauge the amount of water needed; the resultant dilutions tend to be erratic. Rather, the amount of water should be measured with a graduated glass container.

Even if all conditions are standardized, the subsequent mucosal coating can vary from one institution to another. Depending on variations in local water hardness and whether distilled water or tap water is used will influence the quality of mucosal coating. Because of these variables, a number of manufacturers sell premixed liquid preparations. The contrast agent simply is poured into an enema bag without further dilution. **Because with prolonged storage most barium products tend to settle out, vigorous shaking of the barium jugs prior to dispensing is required.** With some products, faster resuspension can be achieved if the jugs are stored on their side rather than upright.

So-called "low-viscosity" barium preparations are available on the market. The manufacturers achieve lower viscosity simply by using less barium sulfate and by other changes in the additives. Unfortunately, some of these products result in a rather thin mucosal coating.

Some radiologists perform colonic lavage prior to the barium enema. Such lavage leads to water retention and subsequent dilution of the barium suspension. The barium manufacturers recognize this factor and market two similar preparations; the one designed to be used following colonic lavage has a barium sulfate suspension of slightly greater specific gravity.

If excessive air bubbles are encountered in a particular locale, empirical addition of a small amount of an antifoam agent should cure the problem.

For optimal results the same commercial barium preparation cannot be used for both the single-contrast and the double-contrast studies. **For the double-contrast study the primary aim is to achieve good mucosal coating. For the single-contrast study, one the other hand, the aim is to achieve homogeneous suspension of the barium particles.** The low concentrations used for the single-contrast examination result in poor mucosal coating.

II. **Toxicity and complications.** Barium sulfate is poorly soluble in water. It is not surprising, however, that a minute portion of any soluble barium would be absorbed from the gastrointestinal tract. Atomic absorption spectrometry shows that following oral ingestion of a commercial barium sulfate preparation, approximately 0.2×10^{-6} of the ingested dose is subsequently excreted via the urinary tract (*Thérapie* 42:239–243, 1987). The reports of such absorption should be viewed as examples of extreme sophistication of available techniques rather than as an indication of any possible toxicity.

A. **Barium aspiration.** Most radiologists believe that aspiration of small amounts of the commercial barium preparations is of little significance. Although initially readily identifiable on radiographs, such barium clears the major bronchi and trachea within hours, leaving little or no residual findings. Aspiration of large amounts, similar to aspiration of any fluid, can result in pneumonia or even significant compromise of pulmonary function.

Following aspiration, most of the barium is readily cleared, although small amounts are ingested by macrophages. This latter residue is generally not visualized on radiographs.

If clinically aspiration is suspected, barium sulfate rather than one of the ionic contrast agents is preferred. Whether the nonionic agents have a role in such a clinical setting is still to be determined.

B. **Allergic reactions.** Hypersensitivity reactions to commercial barium sulfate products have been rare, although within the past several years the number of reported reactions has increased. **Hives, respiratory arrest,** and **anaphylaxis** may develop. Previously, the incidence of reactions was about equal between upper gastrointestinal examinations and barium enemas.

Why the incidence of allergic reactions has increased is not known. Theoretically, barium sulfate should be inert. The commercial barium preparations, however, do contain numerous additives. For many years, such preservatives as the **alkyl *p*-hydroxybenzoates** (such as methylparaben) and similar compounds were used. Because of extensive literature on methylparaben-induced hypersensitivity reactions, commercial barium manufacturers have replaced it by other, more innocuous preservatives. In spite of this, allergic reactions have continued. Lately, most of the reported reactions have been associated with barium enemas.

Allergic reactions have occurred even before the administration of barium sulfate. As a result, such previously thought unlikely candidates as the **latex** used in enema balloons have been implicated. In fact, in 1990 the major manufacturer of barium sulfate products in the United States recalled all enema tips containing latex balloons. Confusing the picture further, latex gloves are routinely used to perform digital rectal examinations prior to the barium enema. Occasional allergic reactions have also been reported following the intravenous injection of **glucagon.**

The etiology of possible allergic reactions during a barium study is still highly speculative. In most patients who do have a reaction, no further testing has been performed to seek out the incriminating agent.

C. **Extraperitoneal perforation.** Most perforations are not reported in the literature. With esophageal perforation and spill of barium into the mediastinum, there is an immediate **inflammatory reaction,** followed by **fibrosis.** Such mediastinal barium can persist for prolonged periods of time, although there is no solid evidence that the sequelae are significantly more severe than with the water-soluble agents. Prior extravasation into soft tissues can be recognized as dense linear opacities.

It is not unusual to study the esophagus for suspected perforation with a water-soluble agent and not see any abnormality. Changing to barium sulfate may allow detection of subtle extravasation.

A survey in Australia and New Zealand found 1 extraperitoneal perforation for every 40,000 barium enemas performed (*Australasian Radiology* 15:140–147, 1971). Most **extraperitoneal perforations** occur in the **rectum** and are not immediately detected. A number of such perforations have been associated with injudicious insufflation of the enema balloon. Barium in the perirectal tissues results in an inflammatory reaction that eventually leads to fibrosis. Such resultant fibrosis can narrow the rectosigmoid and mimic a carcinoma. So far, there is no convincing evidence that barium in the soft tissues is a carcinogen.

Although the water-soluble contrast agents can be used with a chronic or loculated perforation, the higher visibility of barium sulfate will, at times, yield more information. Thus an abscess or other cavity that has been in continuity with the bowel for some time can be studied with barium sulfate.

D. **Barium peritonitis.** The true incidence of barium peritonitis is not known. The **mortality is significant** and in some reviews has approached 50%. Even sterile barium in the peritoneal cavity leads to peritonitis. It induces a concentration of leukocytes here, the barium crystals become coated by a fibrin membrane, and there is a marked in-pouring of fluid into the peritoneal cavity. The introduction of **bacteria** during the perforation can lead to **sepsis** and **shock** within hours. If this goes untreated, the massive in-pouring of fluid can result in **marked hypovolemia.**

The immediate management of barium peritonitis consists of **large volumes of intravenous fluids;** the amounts required are considerably greater than simple maintenance levels. **Antibiotic therapy** is common because of possible bacterial contamination. Most of these patients undergo **surgery,** where the site of perforation can be closed and some of the barium can be evacuated from the peritoneal cavity. Invariably, there is significant residual barium that can be subsequently appreciated by either conventional radiology or computed tomography. The barium crystals embed on the peritoneal surface and resist removal.

If the patient survives the acute episode of barium peritonitis, extensive **fibrosis** and **granulomatous formation** result. The dense fibrosis can involve adjacent structures; for instance, subsequent ureteral obstruction has been reported.

Chapter / 17
Water-Soluble Gastrointestinal Contrast Agents

Jovitas Skucas

Considerable controversy developed during the past several decades concerning the relative indications and contraindications of water-soluble gastrointestinal agents and when they are preferred to barium sulfate. These agents do not coat the gastrointestinal mucosa; rather, they allow visualization of the intestinal lumen by passive filling.

In general, the water-soluble agents are preferred more by surgeons than radiologists, although exceptions exist. The stimulation of peristalsis in postoperative patients and the lack of radiographically visible sequelae following spill from the gastrointestinal tract are the main reasons why some surgeons prefer them.

The water-soluble agents are used if a perforation is suspected and it is deemed necessary to delineate any underlying abnormality. Likewise, in a setting of possible peritonitis or similar condition where the examination itself may lead to spill from the lumen, the water-soluble agents are preferred. Occasionally, visualization of an abnormality with the water-soluble media is insufficient, and it may be necessary to follow with a barium study (Table 17.1).

I. **Indications**
  A. **Ionic agents.** Because the ionic contrast agents stimulate intestinal peristalsis, faster visualization of distal loops of small bowel can be achieved with their use than with barium sulfate. The need for such faster examinations should be balanced against the decreased X-ray contrast and resolution when these agents are used. In general, best results are obtained in the stomach and proximal small bowel; resultant dilution leads to poor visualization of the ileum. Because of the increased peristalsis, more frequent films should be

Table 17.1. Suggested Use of Oral Contrast Agents

Anatomic Space	"Safe" Contrast Agents
Peritoneal cavity	Nonionic or ionic water-soluble agents
Lung (aspiration)	Barium sulfate suspensions ?Nonionic agents
Pleural space	Either barium sulfate or water-soluble agents
Mediastinum	Either barium sulfate or water-soluble agents
Loculated abscess cavity	Barium sulfate suspensions Water-soluble agents
Suspected intravascular communication	Water-soluble agents

obtained than when a similar study is performed with barium.

B. **Nonionic agents.** The decreased osmolality of the nonionic contrast agents is why they are preferred in certain indications. Ideally, the nonionic agents should be used in those circumstances where a barium study is inappropriate. Their high cost, however, precludes use in most adults. If spillage into the pleura or the peritoneal cavity is suspected, the nonionic agents probably do not offer any significant improvement over their ionic counterparts.

In neonates or young children the nonionic agents are generally preferred if there is a potential for perforation.

If a contrast enema is desired in a neonate with suspected necrotizing enterocolitis, many radiologists use barium as the contrast agent. Because of fluid balance problems, Gastrografin is not commonly employed. For this clinical indication, if there is a risk of perforation, the nonionic contrast agents are often used.

II. **Contraindications.** In young children and in adults with hypovolemia, introducing large amounts of hypertonic agents into the gastrointestinal tract can lead to further **hypovolemia, shock,** and **possible death.** In such a clinical situation, intravascular fluid replacement or, where appropriate, use of a nonionic contrast agent should be considered.

In a setting of **aspiration** or an **esophagotracheal fistula** the **hyperosmolar ionic contrast agents can produce pulmonary edema and death.** The nonionic agents are a reasonable substitute. In most adults, however, judicious use of barium sulfate is preferred.

Chapter / 18
Gastrointestinal Agents in Computed Tomography

Jovitas Skucas

A loop of bowel can mimic a cyst, abscess, or neoplasm. Therefore, in most computed tomography (CT) examinations of the abdomen it is advantageous if a readily identifiable marker outlines the gastrointestinal tract. Over the years, a number of negative, neutral, and positive contrast agents have been used for this task.

I. **Negative contrast agents.** Quite often residual gas within the bowel serves as a marker. This is especially true in the colon. If a nasogastric tube is in place, air can be injected into the stomach and small bowel. With some of the earlier scanners, excessive air led to streak artefacts; these are no longer a problem with most of the modern scanners. With significant amounts of gas, window settings wider than usual are of help.

A number of products containing fat, such as mineral oil, have been used (*Radiology* 164:653–656, 1987). Currently, these have a limited application.

II. **Neutral contrast agents.** Distension of the stomach with water, either by drinking or through a nasogastric tube, allows better visualization of gastric wall thickness. Water is little used in the small bowel; a water-filled loop of bowel can have a similar appearance to that of a cyst or abscess.

III. **Positive contrast agents.** Both iodinated water-soluble agents and various barium sulfate suspensions have been used. The ideal such contrast agent should lead to ready bowel differentiation from surrounding structures without any resultant artefacts. Full-strength suspensions as used in conventional radiography obviously have no role. The resultant high contrast simply leads to streak artefacts.

   A. **Water-soluble agents.** Dilute solutions of the ionic and nonionic water-soluble contrast agents have been used perorally. Their major limitation is poor taste and resultant poor patient acceptance, especially in children and cancer patients. When instilled through a nasogastric tube or as an enema, however, they produce adequate bowel opacification.

Some of the contrast agents available on the market, such as Gastrografin or oral Hypaque, contain **flavoring agents.** These are preferred to the nonflavored agents designed primarily for intravenous use. Some radiologists add additional fruit juice or some other such flavoring to improve their taste.

Most radiologists prefer a 2–5% solution of Gastrografin. At such dilution, the solution is hypoosmolar, yet a number of patients still develop diarrhea following the examination.

If the primary indication for a CT examination is suspected pelvic disease, the patient can drink a contrast agent the evening before the examination. Even 20–30 mL of full-strength Gastrografin can be ingested; subsequent overnight dilution should be sufficient to eliminate most streak artefacts. Opacification of the rectosigmoid is enhanced with ingestion of such full-strength Gastrografin as compared with dilute barium, probably because of the peristalsis-stimulating effect of full-strength Gastrografin.

At the dilutions used, the nonionic contrast agents do not have any real advantage over the ionic contrast agents. In addition, no flavored ionic contrast agent specifically designed for oral use is currently available in the United States.

B. **Barium sulfate.** A full-strength barium sulfate preparation simply cannot be diluted to the low concentrations needed for CT. Invariably following ingestion of such a contrast agent, the barium sulfate particles tend to settle out, leading to inhomogeneous opacification of the bowel; the uppermost portion of a loop of bowel may not contain enough barium for adequate visualization, while the dependent portion is so dense that streak artefacts are produced. The barium manufacturers have addressed this problem by modifying the size of the barium particles and the type and amount of additives included.

Most oral CT products contain a 1–3% barium sulfate suspension; several such products are commercially available. They contain rather small barium sulfate particles that resist settling. The additives are also designed to prevent settling of the barium particles.

Studies have shown "better" patient acceptance of barium contrast agents than of the iodinated products. Both products result in similar bowel opacification. At the low concentrations used, both products do not coat the mucosa but simply fill the bowel lumen. Because of better patient acceptance, many investigators currently prefer the barium sulfate suspensions, except when bowel perforation is suspected.

In study of the thorax it can be useful to outline the esophagus. The typical low-concentration, low-viscosity agents simply do not coat the esophageal mucosa long enough for the entire examination. One option is to have the patient drink small sips prior to each scan. As an alternative, a high-viscosity low-concentration barium paste has been developed for coating the esophagus (*Journal of Computer Assisted Tomography* 9:214–216, 1985). High viscosity is achieved by various nonopaque additives; mucosal adherence is sufficiently prolonged that a typical examination can be completed.

One method of opacifying the stomach and small bowel is to have the patient drink approximately 500 mL of the dilute contrast agent several hours prior to the examination. A similar amount is then ingested immediately before the scan. In most patients the stomach, small bowel, and varying amounts of colon are adequately opacified.

If colonic distension is needed, a dilute positive contrast enema can be administered prior to the CT scan. Generally, 150–250 mL of a contrast agent are sufficient to outline the rectosigmoid. Larger amounts are required if more proximal colon is to be opacified.

Chapter / 19
Fistulographic Contrast Agents

Jovitas Skucas

Study of **cutaneous fistulas** is common in most radiologic practices. Many of these fistulas are secondary to surgery or percutaneous abscess drainage, or they form spontaneously secondary to infection or inflammation such as is seen in Crohn's disease.

A catheter is introduced through the cutaneous opening. With light pressure, the catheter can then be inserted as deeply as possible. A number of **steerable catheters** are available, and these are useful with extensive fistulas, especially if an enterocutaneous fistula is suspected. After initial catheter insertion, contrast is injected to outline the tract or cavity. If further internal communication is suspected, the catheter can then be "steered" in the appropriate direction. A guidewire is useful to direct the catheter is some situations. If needed, appropriate-sized drains can be positioned over a guidewire.

Regardless of catheter or technique used, an **adequate seal** is required at the cutaneous orifice to prevent backflow. Injection of insufficient contrast can result in incomplete filling, inadequate demonstration of an abscess cavity, or failure to visualize an enterocutaneous fistula. The pressure needed to obtain an adequate examination should be individualized. Excessive injection may result in spread of an infection.

Several cutaneous fistulas can be present; they may or may not communicate with each other. In general, all such fistulas should be studied. Any possible internal communication can be appreciated better if a catheter is retained in each fistula as it is studied.

These studies should be performed only under fluoroscopic control. Biplane fluoroscopy is of help in outlining the three-dimensional extent of a fistula.

Occasionally, there is either **intraperitoneal spill** of contrast or **intravasation** (*Radiographic Contrast Agents,* 2nd ed. Gaithersburg, MD: Aspen Publishers, 1989, p. 442). It is because of such a possible complication that the **iodinated water-soluble contrast agents are preferred.** There does not appear to be an appreciable advantage to using the nonionic agents; because of their higher costs, most radiologists generally prefer the ionic contrast agents. A possible exception would be their use in those patients who had a prior significant contrast reaction.

Chapter / 20
Pediatric Contrast Agents

Wilbur L. Smith, Jr.
Edmund A. Franken, Jr.

I. **General considerations.** The radiographic contrast agents used for examination of children do not differ substantially from those used for adults. The low incidence of some disorders, such as myeloma, diabetes, renal failure, and cardiovascular instability, among pediatric patients makes these factors less common problems for contrast selection and examination preparation than in adult populations. **The variation in physical sizes of pediatric patients requires more thorough attention to individualizing dosage.**

The differences in contrast usage between children and adults occur predominantly because of great differences in the spectrum of diagnostic problems between the two age groups. A general outline of contrast indications specific for the pediatric age group is included for each type of study discussed.

II. **Intravascular iodinated contrast agents.** The iodinated contrast materials available for intravenous and intra-arterial use in adults are essentially identical with those for children. These may be considered as the conventional high osmolar contrast media (HOCM), which are triiodinated fully substituted benzene derivatives, and the low osmolar contrast media (LOCM), which are larger molecular dimers exhibiting less osmolar composition per gram of iodine. Both types of contrast have been used extensively in children and are effective for opacification of vascular structures. In general, LOCM are perceived to offer several advantages for pediatric patients, including less discomfort and a lower incidence of adverse reactions (*Radiology* 175:621, 1990); however, to date, a large population-based paired comparison study employing only pediatric patients is not available. The disadvantage of the LOCM is the tremendous cost differential (*New England Journal of Medicine* 323(21):

1463–1468, 1990), which may be as high as a 10-fold increase in expenditure. In 1990, LOCM cost approximately $1/mL, while HOCM cost $.10/mL.

Toxicity from intravascular iodinated contrast in children can be classified as occurring as a result of three mechanisms: (1) **chemical toxicity, (2) osmolar toxicity,** and **(3) anaphylactoid-type reactions.** (J. Skucas, ed. *Radiographic Contrast Agents*. Gaithersburg, MD: Aspen Publishers, 1989, pp. 462–485).

Chemical toxicity is most frequently seen in neonates. **Neurotoxicity,** usually manifest as seizures, occurs because of chemical effects from intravascular contrast and is **most likely to occur in neonates with preexisting neurologic disorders** (*AJNR* 7:522–525, 1986). While there is a suggestion that this class of adverse reaction may be less frequent with LOCM, the data to support this are not definitive. **Pulmonary toxicity,** manifest as **pulmonary hemorrhage** or **pulmonary edema,** may also occur either due to direct toxicity or in overdose situations (*Radiology* 139:337–339, 1981). Contrast nephrotoxicity occurs in children and adults. The mechanism of this disorder is unclear. Initial work suggested that the osmolality of the contrast played a major role in creating an "osmolar nephrosis"; however, the disorder also occurs with LOCM administration. It is unclear if the incidence of osmolar nephrosis is the same between LOCM and HOCM (*Radiology* 167:607–611, 1988).

Contrast accidents with acute osmolar toxicity as a result of intravascular iodinated contrast overdose occur more frequently in children than in adults. Failure to recognize the extreme variability in body mass probably accounts for most of these mishaps (*New York Journal of Medicine* 73:1958–1966, 1973). Contrast loads in excess of **3 mg/kg body weight** are likely to cause rapid and substantial fluid shifts with hemoconcentration and depression of cardiac output. **Seizures, cardiac decompensation, and death** have been reported as a result of acute osmolar poisoning in children. As with contrast toxicity, neonates are most susceptible. Table 20.1 gives an outline of the osmolality of the commonly used intravascular contrast media for pediatric patients.

Idiosyncratic reaction, the "contrast allergy reaction," occurs infrequently in children. A review of the experience of members of the Society for Pediatric Radiology documented only 5 severe reactions and no deaths in a population of 12,000 children undergoing intravenous urography (*AJR* 123:802, 1975). All of the patients studied in that review received HOCM. A more

Table 20.1. Approximate Osmolality of Commonly Used Intravenous Contrasts

Contrast	Trade Name	Iodine (mg/mL)	Osmolality (mOsm/kg)
Diatrizoate	Hypaque 60	280	1500
	Reno-M-60	280	1500
Iothalamate	Conray-60	280	1500
Iohexol	Omnipaque 300	300	700
Iopamidol	Isovue 300	300	600
Ioxaglate	Hexabrix	320	600

recent study documented that the **highest incidence of adverse contrast reactions was in the 10–19-year-old range** and that the incidence of adverse reactions was significantly lower among those receiving LOCM (*Radiology* 175:621–628, 1989). Similar trends toward less adverse reactions were noted in the 1–9-year-old group. The numbers were insufficient to document the trend in the children younger than 1 year.

The issue of the appropriate intravascular contrast for children is difficult. The results suggest that LOCM have a lower incidence of adverse drug reactions and are more comfortable when injected. The latter is not a trivial fact when dealing with a lightly sedated child where cooperation is needed. On the other hand, none of the studies, to date, have documented excess mortality in the pediatric age group when appropriate doses of HOCM were employed. The cost differential for LOCM is not trivial (*New England Journal of Medicine* (21):1463–1468, 1990). These competing factors have led many practitioners to use a selection criterion for use of LOCM. In our present practice we use LOCM (Table 20.2) for **neonates, sedated children, cardiac studies, and those children with prior history of contrast reaction.** With these screening criteria, approximately 50% of our patients now receive LOCM; however, this proportion is increasing steadily.

III. **Excretory urography.** The number of excretory urograms performed in children is considerably lower than 10 years ago because of the substitution of other imaging modalities, particularly ultrasound and nuclear medicine examinations. The technique is still valuable for renal calculi and for anatomic delineation of the renal pelvis and ureters in selected cases.

In children, good excretory urograms can be obtained with a bolus IV injection of iodinated contrast material of **1 mL/kg, (up to a maximum of 50 mL).** This is followed by an early (1–2

Table 20.2. Selection for LOCM

Neonates
Sedated children
Cardiac studies
Children with a history of prior reaction to contrast media

minutes after bolus completion) radiograph to document the nephrogram phase, followed by delayed radiographs to delineate the collecting systems. In children, the upper tracts can be spuriously dilated as a result of a distended urinary bladder, and in this situation we obtain a postvoid radiograph. In preparation for the urogram, dehydration of the child is usually not necessary or appropriate. As nausea is a common side effect of intravascular contrast administration (especially with HOCM) and **as a general precaution to prevent vomiting we require that our patients have nothing by mouth for 4 hours prior to the study.**

IV. **Angiocardiography and angiography.** Contrast administration for angiocardiography and angiography must adhere carefully to the osmolar limitation guidelines. As multiple injections of contrast are frequent, studies must be carefully planned so that a dose limitation of **3 mL/kg** body weight is adhered to while sufficient diagnostic information is obtained. Intracardiac injection of conventional HOCM has been documented to cause physiologic derangements including **decreased cardiac contractility, changes in cardiac conductivity** (as measured by electrocardiogram), **decrease in stroke volume,** and **increase in heart rate** (*Radiology* 148:687–691, 1983). LOCM cause less pronounced effects and have achieved widespread usage in angiocardiography for this reason. While some of the physiologic effects of intra-arterial or intracardiac HOCM may be obviated by the use of LOCM, several studies suggest that the nephrotoxic effects of the two contrasts are similar (*Radiology* 167:607–611, 1988).

V. **Computed tomography.** Intravenous contrast enhancement for computed tomography (CT) in children is an invaluable adjunct to the study. In general, most abdominal CT studies, any chest study where the mediastinum or great vessels are the focus, and selected brain CT studies are performed with contrast. A table of general indications for choosing to administer contrast for CT is given in Table 20.3.

Except in special circumstances, we administer a bolus injection of contrast material followed by rapid sequence scan-

Table 20.3. Indications for Contrast Administration for CT

Cranial CT		
With contrast		*Without contrast*
Tumors		Trauma
Infection (beyond neonate)		Neonatal infection
		Anomalies
Abdominal CT		
With IV contrast		*With GI^a contrast*
Tumors		Inflammatory bowel disease
Trauma		
Abscess		Bowel tumor
Anomalies, especially renal		Abscess outside solid *organs*
Pelvic CT		
With IV contrast	*With GI^a contrast*	*Without contrast*
Trauma	Tumor	Bony abnormalities
Tumor	Abscess	
Abscess	Anomalies	
Pelvic anomalies		
Musculoskeletal		
With IV contrast		
Tumors of bone and soft tissue		
Soft tissue abscess		

^aGastrointestinal.

ning rather than an IV drip. In general, a **dose of 1 mL/pound up to a maximum of 50 mL** is adequate for most pediatric patients. Changes in scanner technology allowing shorter scan times have facilitated and expanded the usefulness of bolus scanning. Pediatric patients are often sedated for CT studies, and we use LOCM for these patients because of their lack of local irritation and pain on bolus injection. The higher contrast cost is a trade-off for the improved patient throughput and lack of patient motion resulting in better scans.

I. Intravenous contrast for magnetic resonance imaging. The indications for contrast enhancement in magnetic resonance imaging (MRI) of children are still in the developmental stage. The initial Food and Drug Administration approval for use of the drug for evaluation of the central nervous system in children over age 2 years was granted in 1989. Several studies suggest that there is substantial improvement in detection of tumors of the central nervous system in children after contrast administration (*Radiology* 169:123–126, 1988). Anecdotal information indicates contrast enhancement to be of value in both tumor and inflammatory conditions of the meninges, some abdominal tumors, and inflammatory conditions of the brain. Further development of indications for contrast enhancement in MRI will evolve as experience becomes more widespread.

The usual MRI contrast employed for children is gadolinium diethylenetriamine pentaacetic acid (Gd-DTPA). The active component of this compound is the gadolinium cation that contains unpaired orbital electrons, allowing it to alter a magnetic field. The DTPA acts as a carrier for the gadolinium, and the compound generally follows the distribution of extracellular water in the body. Lesion enhancement occurs due to the presence of extra free water in the area of the abnormal focus. In the case of the brain this is related to the local breakdown of the blood-brain barrier at the site of a lesion (*Radiology* 176:225–230, 1990).

DTPA is eliminated almost exclusively in the urine. The clearance rate of DTPA is subject to the rate of glomerular filtration. Impaired renal function is, therefore, a relative contraindication to gadolinium administration (*AJR* 142:625–630, 1984).

The usual intravenous dosage of Gd-DTPA is **0.1 mmol/kg** IV, the same as the adult dose. Children older than 6 weeks of age essentially handle gadolinium over the same time course as an adult, so that optimal scanning takes place within a few minutes after the injection of the contrast. Neonates are an important exception to this rule. Neonates have a relatively low glomerular filtration rate, and therefore, the elimination of gadolinium is delayed, allowing for prolonged scanning times after the administration of gadolinium. This feature can occasionally work as a benefit to the imager, as adequate scans can be obtained as long as 4 hours after the injection of gadolinium to the neonate (*Radiology* 176:225–230, 1990).

Serious adverse reactions to gadolinium are unusual, and no separate report of adverse gadolinium reactions in children is identified. There are, at present, several reports of anaphylactoid reactions to gadolinium (*AJNR* 11:1167, 1990); however, the incidence of such reactions, while lower than that for iodinated contrast, is uncertain (*AJNR* 11:1168, 1990). Local reactions occur in as many as 20% of patients and include pain or a feeling of "coldness" at the injection site, headaches, and nausea. Contraindications to the use of gadolinium include prior reactions and severe renal compromise.

VII. Gastrointestinal tract contrasts. The choices of gastrointestinal tract contrast for children are tailored to the type of examination desired. In general, single-contrast studies of the gastrointestinal tract are performed in infants and neonates where congenital anomalies are most likely and double-contrast studies of the alimentary tract are reserved for those children suspected of having specific mucosal abnormalities such as juvenile polyps.

In the neonate, air represents an extremely safe and effective contrast material for evaluation of many congenital gastrointestinal abnormalities. The insufflation of 20–30 mL of air into the stomach is usually sufficient to make the diagnosis of duodenal atresia or proximal small bowel obstruction or to opacify the bowel to clarify a diagnosis of congenital diaphragmatic hernia presenting as opaque hemithorax. In cases of suspected esophageal atresia where the nasogastric tube cannot be passed, insufflation of a smaller amount of air, usually 5–10 mL, is sufficient to make the diagnosis and to differentiate this condition from esophageal perforation. Air is an extremely safe contrast, with the only hazard being overdistension of a viscus proximal to a perforation. As long as one adheres to a limit of 30 mL within the stomach and distal bowel or 10 mL within the proximal pouch of an esophageal atresia, this risk is minimal (J. Skucas, ed. *Radiographic Contrast Agents*. Gaithersburg, MD: Aspen Publishers, 1989, pp. 462–485).

Barium preparations are generally the contrast of choice for upper gastrointestinal, small bowel, and colon studies. **Barium is contraindicated in those situations where perforation of a hollow viscus is likely.** Data from the 1950s suggest that the presence of barium in the peritoneal cavity leads to a higher mortality than that of the spilled visceral contents alone. More recent work suggests that with the current methods of patient support this may no longer be valid (*AJR* 145:665–669, 1985); however, when given a choice it is prudent not to use barium in the face of probable gastrointestinal perforation. **The second potential hazard of barium is hypotonicity.** In neonates and those infants with motility disorders such as Hirschsprung's disease, retention of large amounts of barium suspensions may result in absorption of water from the suspension and water intoxication (*Gastrointestinal Imaging in Pediatrics*. Hagerstown, MD: Harper & Row, 1975). This problem is readily circumvented by the use of normal saline to mix the barium or by the addition of 1 tablespoon of sodium chloride per gallon of barium suspension when barium is used for these patients. Rare instances of allergic reactions after barium administration have been reported; however, these are thought usually to be due to the flavoring or carrier with which the barium is suspended. There have also been anecdotal reports of aggravation of obstructive conditions by barium inspissation (*Pediatric Radiology* 14:230, 1984); however, this is such an infrequent event that it should not have a major impact on the choice of contrast material.

The weight-to-volume barium employed for single-contrast studies of the upper gastrointestinal tract in children ranges from 50% to 100%. For single-contrast colon examination, 15–25% weight-to-volume barium suspensions suffice. Air contrast studies employ a denser barium using 200–250% weight-to-volume for the upper gastrointestinal tract and 80% weight-to-volume for double-contrast colon studies. Small bowel follow-through barium suspensions are identical with those used for the single-contrast upper gastrointestinal tract while enteroclysis studies generally use an approximately 30% weight-to-volume suspension. (*Critical Reviews in Diagnostic Imaging* 30(4):317–340, 1990).

In general, barium studies of the upper gastrointestinal tract in neonates are best undertaken by employing a nasogastric tube and small amounts of contrast, followed by a larger amount of air to distend the stomach until the anatomy of the stomach, duodenum and duodenal jejunal junction are delineated. We generally use approximately 10 mL of barium and 30–40 mL of air to accomplish this. Thereafter, larger amounts of barium can be instilled into the stomach to distend it and to study the distended duodenal bulb and duodenum. For studies to evaluate esophageal reflux it is customary to withdraw the nasogastric tube before maneuvers to elicit reflux are begun. The appropriate maneuvers to elicit gastroesophageal reflux are of some controversy. We tend to be minimalist in the maneuvers, using predominantly postural change to elicit the reflux.

In older children the gastrointestinal series can continue in the same fashion as would be performed in an adult; however, less emphasis is usually placed on the detection of ulcer disease. Owing to the not infrequent presentation of Crohn's disease as vague abdominal pain, we often obtain delayed films in older children with an appropriate history to evaluate the small bowel for the possibility of Crohn's disease mimicking ulcer disease.

Contrast employed for CT scanning is either a 1% weight-to-volume solution of barium or iodinated contrast material. Owing to the low body fat content of children, especially infants, we find that more satisfactory bowel opacification can be obtained by diluting these standard adult contrasts with an equal volume of water. We usually administer the oral contrast 1–2 hours before the scan, allowing bowel transit to opacify the colon. An additional oral bolus is administered just before scanning in order to opacify the stomach and duodenum. For pelvic studies we administer a contrast enema with the same contrast dilution used.

Water-soluble contrast material is occasionally of value in evaluation of the gastrointestinal tract, particularly in those instances where perforation of fistula formation is suspected. Two general types of water-soluble contrast are available, the HOCM, which use the traditional triiodinated organic compounds, and the LOCM, which are functionally identical with the LOCM used intravenously.

HOCM must be used with care and have several specific disadvantages. **They are extremely toxic to the lungs and can cause pulmonary edema if aspirated.** They are, therefore, contraindicated in vomiting infants or other situations where their passage into the tracheobronchial tree is likely. **HOCM also cause direct mucosal damage as a result of the osmolar effect on the gastrointestinal lining cells. It is, therefore, not advisable to instill them in a closed loop or an obstructed area** (*Radiology* 122:693–696, 1976). HOCM draws water into the bowel lumen, causing dilution of the contrast, and also in infants, creating the possibility of dehydration. This property of drawing water by osmosis is of value is some specific instances, such as the treatment of meconium ileus and meconium ileus equivalent (see discussion later in this chapter under therapeutic use of contrast). This property also makes HOCM inappropriate for study of small bowel lesions, as the contrast is usually so diluted by the time it reaches the distal small bowel that diagnosis is hindered.

LOCM do not have the disadvantages of causing fluid shifts or drawing water into the bowel. They can, therefore, be used for a small bowel study as well as for studies of the stomach, duodenum, and colon. LOCM give poor muscosal coating and generally are inferior to barium for evaluation of muscosal disease (*Clinical Radiology* 31:635–641, 1980). LOCM are relatively safe if accidental aspiration should occur (*Pediatric Radiology* 14:158–160, 1984). There is some indication that LOCM may cause some mucosal cellular alteration; however, the degree of this is substantially less than with HOCM. The major drawback of LOCM is the cost, especially if large volumes are used. In general, it is our recommendation for neonates with suspected perforation that LOCM be employed, as most studies can be accomplished with as little as 20 mL of contrast material, the cost of which is insignificant relative to the costs of possible complications.

VIII. **Cystourethrography.** Cystourethrography, particularly with a voiding sequence (voiding cystourethrography (VCU)), is an important method of evaluating the lower urinary tract in children. The procedure is essential to assess structural and functional abnormalities in the child with suspected obstruction of the lower urinary tract. It is currently the most popular method for assessing vesicoureteral reflux (VUR), an important component of the evaluation of the child with urinary tract infection. The VCU is reasonably reliable in detecting VUR but of less value in grading the reflux (*AJR* 153:807–810, 1989).

Alternate methods of study exist, and radionuclide cystography should be considered for evaluation of VUR, particularly for follow-up studies, as radiation dose is substantially reduced. Ultrasound supplemented with Doppler of the "ureteral jet" has an advantage in study of the lower urinary tract, in that catheterization with subsequent risk of sepsis can be avoided; however, its reliability is still under investigation.

A variety of radiopague agents for VCU have been recommended over the years. In general, most VCUs in North America are performed by using the traditional agents diatrizoate (Hypaque, Renografin) or iothalamate (Cysto-conray). There is no substantive evidence at this time that the lower osmolar agents are less toxic to the bladder for these studies.

Concentration of the contrast agent is of more importance than the specific agent selected. Available packaged units generally contain 25–30% iodine. Considerably less concentration (10% is isotonic) can be utilized with a much lower anticipated rate of complications (*Journal of Urology* 128:1006–1008, 1982).

The amount of contrast to be administered for VCU varies according to the age of the patient, information desired, and status of the urinary tract. Avoidance of overdosage of the child with a massively dilated urinary tract must be avoided. In the usual circumstances of a normal bladder in a child with urinary tract infection, sufficient volume must be given to assure prompt micturition. Clinical judgment of the attending radiologist is necessary in this situation, since infants are unable to indicate their need to void.

Techniques of administration are probably of greater importance than the agent used to keep the number of complications low (*Radiology* 155:105–106, 1985). This includes strict aseptic technique infusion by gravity drip. **Local complications include local reactions within the bladder, such as cystitis.** Rarely, systemic complications occur due to absorption of contrast material for the bladder.

In summary, cystourethrography in children can be performed by using any of the currently popular water-soluble agents. **A concentration of 10% is suggested as offering optimal safety yet diagnostic images.** Strict adherence to technique is the most important segment of the examination.

IX. **Bronchography.** Bronchography is seldom utilized in North America or Western Europe in infants and children. This is for several reasons: (1) bronchiectasis is an infrequent disease; (2) alternate diagnostic methods, particularly CT and endoscopy, are superior (*Annales de Radiologie (Paris)* 31:25–33, 1988); and (3) complications of the procedure are frequent in infants and children.

Possible **indications** for bronchography in children include the **detection of congenital anomalies, further characterization of abnormalities found at bronchoscopy, and demonstration of fistulae** (*Pediatric Radiology* 14:158–160, 1984). The availability of flexible endoscopy and thin-section or dynamic CT scanning obviates many of the indications for bronchography in children.

All agents used for bronchography have two problems: (1) toxic effect on the bronchi and pulmonary tissue, and (2) mechanical effects in the airway. Mechanical airway obstruction is a particular problem in infants because of the small size of the airways. Barium has been utilized in the past for bronchography. Most available barium mixtures in current use contain multiple additives, many of which are potentially toxic to the bronchi and lungs. **The classical intravascular water-soluble agents should never be used for bronchography, as they produce pulmonary edema and an intense inflammatory response.** Oil-based agents such as dionosil have a long history of use in pediatric bronchography. Our recommendations for this agent are 1.25 mL/year of age for the right lung and 1.0 mL/year of age for the left lung. (J. Skucas, ed. *Radiographic Contrast Agents.* Gaithersburg, MD: Aspen Publishers, 1989, pp. 480). The low osmolar agents have been used for bronchography in infants and children, with mixed results. Metrizamide in a concentration of 220 mg iodine per mL has been clinically without evident harm (*Pediatric Radiology* 14:158–160, 1984). Nevertheless, there is some experimental evidence that similar agents damage tracheal epithelium (*Pediatric Radiology* 10: 440–443, 1990).

Most bronchograms in infants and children are performed with the patient under general anesthesia because of inability of the child to cooperate. This creates some technical problems, in that the usual spontaneous respiratory efforts to distribute bronchographic contrast are substituted by the positive pressure application of the anesthetist. Segmental or lobar collapse is common, and fatalities have been noted.

In the infrequent instances where pediatric bronchography is required, dionosil is probably the agent of choice in older infants and children. In the rare instances where examination is needed in the neonate or infant, low osmolar agents are preferred because of lack of mechanical airway obstruction.

X. **Biliary agents.** We have now gone 10 years in pediatric radiology section of this department without performing an oral cholecystogram or classical intravenous cholangiogram. Alternate methods of study of the biliary tract, particularly ultrasound, are of much greater value than the more antiquated methods. Contrast agents used for endoscopic retrograde cholangiopancreatography (ERCP) and percutaneous transhepatic cholangiography are the same used in adults. In the unusual instance where such studies are required, the following oral dosages of Telepaque are employed: for patients less than 13 kg body weight, 0.15 gm/kg; for those 13–25 kg body weight, 2 gm; and for those over 25 kg body weight, 3 gm.

XI. **Therapeutic use of contrast material**
   A. **Treatment of meconium ileus**
      1. **Indications.** Infants with intestinal obstruction owing to meconium ileus can avoid laparotomy if enemas with water-soluble agents are administered. The overwhelming numbers of neonates with meconium ileus have cystic fibrosis; prognosis is this group is considerably improved when this year is compared with previous years (*Archives of Surgery* 124:837–840, 1989).
      2. **Pathologic nature of meconium ileus.** Affected infants have intestinal obstruction at birth because of the presence of an abnormal meconium filling the distal ileum and, to a lesser extent, the colon. Such neonates show evidence of intestinal obstruction on plain film. Contrast

enema shows a microcolon, along with a dilated ileum filled with intraluminal material.
3. **Therapeutic contrast enemas.** For many years it has been recognized that after enemas with water-soluble contrast material, the infant with meconium ileus might pass the meconium per rectum, with subsequent alleviation of intestinal obstruction. Mechanisms for this are not entirely clear. Postulated causes are **(1) surface wetting effect of Gastrografin (Tween-80), (2) mechanical effects of the enema, and (3) osmotic diarrhea induced by hypertonic contrast material.**
4. **Complications.** Perforation is a theoretical complication of any contrast enema in the neonate. Utilizing hypertonic contrast material creates a "third space" effect with an infusion of body fluids into the colon and ileum. At the same time, hypertonic contrast material is absorbed systemically and induces diuresis. The end result is reduced blood volume and cardiac output with potential for shock (*Anesthesiology* 61:454–456, 1984). Because of the hazards of the procedure, coordination with the surgeon and neonatologist is essential.
5. **Administration of therapeutic enema in meconium ileus.** Consensus indicates that there is a need to produce an osmotic diarrhea, refluxing contrast material into the inspissated material in the dilated ileum. Although Gastrografin may be used, our own preference is **Hypaque in a concentration of 25–40%.** This should be of somewhat less toxicity than Gastrografin. The technique of the procedure consists of filling the colon with reflux of contrast material into the dilated ileum. Periodic films are made over the next several hours to monitor passage of the meconium.
6. **Other uses for water-soluble contrast agents in the gut.** Meconium ileus equivalent is that situation in older patients with cystic fibrosis in whom intestinal obstruction develops in the ileum because of insufficient digestion of food. Oral and rectal contrast can be used to relieve this condition as in meconium ileus. Similarly, liquefaction of stool in severely constipated children may be accomplished with water-soluble contrast agents.
7. **Summary.** Therapeutic contrast enema with water-soluble agents is of considerable benefit to the neonate

with meconium ileus and allied conditions. As the procedure is quite hazardous, **it is recommended only in those circumstances where consultation with a pediatric surgeon and a neonatologist is available.**

XII. Intussusception
A. Barium enema reduction
1. **Reduction of intussusception** by therapeutic enema has been known for over 100 years. Indications and techniques for using barium enema to reduce intussusception are well established. This procedure has the advantage of use for many years. In skilled hands it is efficacious, with reduction of intussusception in most cases (*Pediatric Radiology* 20:57–60, 1989). **Evidence of perforation or peritonitis are contraindications for the reduction of intussusception.** Small bowel obstruction, prolonged duration of signs and symptoms, and location of the intussusception in the left colon or rectum predict less likelihood of success for enema reduction. The interested reader is referred to the above article for details on hydrostatic reduction.
2. **Water-soluble contrast agents in reduction of intussusception.** Use of water-soluble contrast agents in reduction of intussusception is advocated by some because of the potential risk of colon perforation in reduction of intussusception. Majority opinion would indicate that the risk of complications from water-soluble agents probably outweighs the danger of additional complications if perforation occurs.
3. **Pneumatic reduction of intussusception.** Pneumatic reduction of intussusception has recently become popular in North America and Australia, having been performed in China and Argentina for many years (*AJR* 150:1345–1354, 1988; *Pediatric Radiology* 20:472–477, 1990). Techniques of performance of pneumatic reduction are found in the above references. In general, air at a pressure of 80–120 mm Hg is used to reduce the intussusception. Preliminary experience indicates that this is a safe procedure, performed more quickly than barium reduction and with a comparable success rate. Gas entering the terminal ileum during this procedure is not necessarily a sign of successful reduction (*Radiology* 174:187–189, 1990).

4. **Summary.** For those physicians associated with institutions where controlled studies are in progress on gas reduction of intussusception or where the radiology team has the appropriate expertise and equipment to attempt gas reduction, this might be considered. In most other instances, barium enema reduction can be performed without additional morbidity or mortality.

INDEX

Page numbers followed by *italic* "t" denote tables; those followed by *italic* "f" denote figures.

Abdomen
  dynamic scanning of, 77
  indications for computed tomography in, 66
  infusion technique for surgery patients, 61
  plain film of, 36
  routine CT protocol for, 78
Abdominal aorta
  aneurysm of, 58
  visualization of, 94
Acidosis, dialysis for, 33
Acorn-tip cannula, 133
Acromion process, 122
Activation systems, contrast media effects on, 173–175
Acute lethal dose (LD_{50}), 8
  reduction of, 34
Acute renal failure
  causes of, 38–39
  contrast-induced, 28–35
  ultrasonographic diagnosis of, 38
Adenomas, adrenal gland, 158
Adrenal glands, magnetic resonance scanning of, 158
Adrenergic agonists, for contrast reactions, 177
Adrenergic drugs, for bronchospastic reaction, 24
Air
  as safe, effective GI contrast in children, 213
  as second contrast agent, 189–190
Allergic reactions. *See also* Anaphylactic/anaphylactoid reactions
  to barium sulfate, 197–198
  with hysterosalpingography, 137–138
  to iodinated contrast medium, 92
  to nonionic contrast media, 93
Allergy, and risk of reactions to contrast media, 166, 167

Alpha-agonists, to treat contrast reactions, 178
AMI-121, 150
AMI-25, 149
Amipaque
  contraindications for, 113–114
  CSF leakage with, 111
  development of, 84
  drawbacks and advantages of, 109
  introduction of, 107
  lowered seizure threshold and, 113–114
  neurotoxicity of, 112
  toxicity of, 112–113
Anaphylactic/anaphylactoid reactions. *See also* Allergic reactions
  in children, 208–209
  as contraindication for MRI, 160
  to gadopentetate dimeglumine, 151
  identification of, 22
  treatment of, 24
  to venography, 105–106
Anaphylatoxins, activation of, 173
Anaphylaxis, immunoglobulin E in, 173
Aneurysms, detection of, 88
Angiocardiography, in children, 210
Angiography, 1. *See also* Digital subtraction angiography; Neuroangiography
  for children, 210
  complications with, 97–98
    characterization of, 98–100
    risk factors and prevention of, 100–101
  contrast media injection technique in, 91–92
  digital subtraction, 96–97
  future considerations for, 101
  general considerations with, 94

Angiography—*continued*
  hepatic computed tomographic, 69–70
  indications for use of low-osmolality contrast agents in, 101*t*
  magnetic resonance, 154
  peripheral, 95–96
  in secondary azotemia diagnosis, 49
  visceral, 94–95
Ankle
  arthrographic technique for, 124–126
  films of, 125–126
  lateral line drawing of, 125*f*
Antecubital vein route, 59
Antegrade pyelography, 42–43
  for azotemia primary diagnosis, 46
Antegrade small bowel examination, 193
Anticonvulsants, for seizure reaction, 27
Antigen-modulated hypersensitivity, 172
Antihistamines
  to prevent reactions, 170
  to treat contrast reactions, 24, 178
Anxiety, role of in contrast media reaction, 168, 175
Arachidonic acid
  activation of, 175
  mobilization of, 177
Arachnoiditis
  as contraindication for myelography, 114
  with oil-based contrast media, 108
Arrhythmia, in contrast media reaction, 99
Arterial portography, 69
Arteriography, for azotemia diagnosis, 43, 45
Arteriovenous malformations, spinal, 156
Arteritis, neuroangiography for, 90
Arthrography
  complications of, 117–118
  contraindications for, 117
  indications for, 117
  sterile preparation and local anesthesia for, 118–119
  technique of
    for ankle, 124–126
    for elbow, 126
    for hip, 122–124
    for knees, 119–120
    for shoulder, 120–122
    for temporomandibular joint, 128–130
    for wrist, 126–128
Asthma, and risk of reactions to contrast media, 166
Astrocytomas, 155

Atropine
  for treatment of reactions, 170
  for vagal reactions, 25
Autonomic nervous system, in reactions to contrast media, 168
Azotemia
  without collecting system obstruction, 44–46
  with collecting system obstruction, 46–47
  causes of, 40*t*
  diagnostic tools for, 36–44
  disease states associated with, 38–39
  general causes of, 44
  imaging with, 36–49
  primary diagnostic pathways for, 44–47
  secondary diagnostic pathways for, 45*f*, 47–49

Balloon catheter, used in hysterosalpingography, 133, 134*f*
Barium
  aspiration of, 197
  contraindications and potential hazards of for children, 213–214
  inspissation of, 214
  in upper gastrointestinal tract studies in neonates, 214
  weight-to-volume for children, 214
Barium enema
  double-contrast, 196–197
  for reduction of intussusception in children, 220, 221
  single-contrast, 195–196
Barium paste, 192
Barium peritonitis, 199
Barium sulfate
  allergic reactions to, 197–198
  aspiration of, 197
  basic properties of, 187–189
  carbon dioxide added to, 190
  clinical application of, 187–197
  in colon studies, 195–197
  commercial preparations of, 189
  effervescent, 190–191
  extraperitoneal perforation with, 198–198
  for gastrointestinal studies, 201
    computed tomography, 203–204
    in small bowel, 193–195
    of upper tract, 191–192
  ideal suspension of, 188–189
  large-particle, high-density, 188
  particles of, 187
  suspension of, 187–188
  toxicity and complications of, 197–199
  viscosity of, 188, 189

weight of, 189
Barium sulfate tablets, 192
Bedrest, postmyelography, 114–115
Benzoic acid molecule, triiodinated, 83
Beta-2-agonist inhaler, 24
Beta-agonists
 to prevent reactions, 177
 to treat contrast reactions, 178
Biexponential decay curve, 8
Biliary agents, in children, 218
Biliary tree, computed tomographic protocol for, 78
Biopsy, for azotemia diagnosis, 43–44
 primary, 44–45
 secondary, 49
Blood
 coagulation of and contrast media, 86–87
 contrast media pathophysiology in, 84–87
 viscosity of, 15, 85
Blood-brain barrier (BBB)
 in central nervous system function, 110
 defect of in area postrema, 169
 diseases causing damage to, 88–89
 disruption of, 152, 153
 function of, 88
 integrity of, 16
 reduced effect on with low osmolar agents, 164
 risk for damage to, 93
Blood-contrast media mixing
 high osmolality of, 15
 and injection rate, 1–14
Bolus injection technique, 68–69
 for intravenous DSA and dynamic CT, 11
Bowels
 MRI scanning of, 158
 opacification of in adults vs. children, 214–215
Bradykinin
 activation of, 174, 175
 generation of in contrast reaction pathway, 21
 in mobilization of arachidonic acid, 177
 stimulation of production of, 18
Brain
 contrast media effects with metastases to, 16
 detection of parenchymal infection in, 153
 identification of tumor in, 152
Breast cancer, 157
Breast imaging, 157
Breath-hold fast imaging techniques, 157–158
Bronchogenic carcinoma, 157

Bronchography, in children, 217–218
 contrast agents used for, 217–218
 indications for, 217
Bronchospasm
 with high vs. low osmolar agents, 164
 identification of, 22
 mechanism of in reaction to contrast media, 169
 treatment of, 23–24
Bulk susceptibility effects, 143, 145

Calcium disodium-EDTA additive, 4
Calculus disease, evaluation of, 79
Cannulas, used in hysterosalpingography, 133
Cardia arrhythmia/arrest, mechanism of in contrast reaction, 169
Cardiac decompensation, in children, 208
Cardiac disease, 100
Cardioversion, for treatment of reactions, 170
Catheter
 insertion of, 205
 percutaneous placement of, 94
Celiac axis, visualization of, 94, 95
Central nervous system (CNS)
 blood vessels of, 87–89
 contrast media penetration into, 16
 as mediator in contrast reactions, 21–22
 physiology of and neurotoxicity with water-soluble contrast media, 109–111
 protection of, 110
 role of in contrast reaction, 167–170
 vascular contrast media use in, 83–93
Cerebral infarction, detection of, 153–154
Cerebritis, 153
Cerebrospinal fluid (CSF)
 composition of, 110
 flow of, 111
 leakage of, 111
 production and drainage of, 111, 113
  factors reducing, 111t
 volume of, 110–111
Cerebrovascular disease, and contrast media reactions, 100
Chemical synovitis, 117
Chemotoxic reactions, 171
 in children, 208
Chest, routine CT protocol for, 79
Cholinesterase, inhibition of, 164–165
Cimetidine (Tagamet)
 for anaphylactoid reaction, 24

INDEX

Cimetidine—*continued*
  for urticaria, 23
Clearance, of X-ray contrast media, 8–10
CM-ARF. *See* Contrast medium-induced acute renal failure
Coagulation, retardation of, 98
Coagulation system activation, 173
  in contrast reaction pathway, 21
Collecting system obstruction
  azotemia diagnosis with, 46–47
  causes of chronic azotemia with, 40t
  causes of false negative ultrasonographic diagnoses for, 37t
  extraluminal, 47
  intraluminal, 46
  pyelography for, 42–43
  ultrasonographic diagnosis with, 38
Colon
  barium studies of, 195–197
  distension of, 204
Color Doppler ultrasonography, 45
Complement activation
  contrast media effects on, 18, 165, 173–174
  in contrast reactions, 21, 167, 176–177
Computed tomographic portography, 69
  for liver evaluation, 78
Computed tomography (CT)
  in azotemia diagnosis, 40–41
    primary, 45–47
    secondary, 47–49
  for children, 210–211
  contrast media use in, 66–82
  false negative diagnoses with, 40
  false positive diagnoses with, 40
  gastrointestinal agents in, 202–204
    in children, 214–215
  of head and body, 1
  with intravenous contrast media, 40
  vs. myelography, 115
  postmyelography, 115, 116
Concentration, chemical measures of, 4–6
Congestive heart failure, and contrast media reactions, 100
Conray 43, 97
Consent, informed, 185–186
Consumption coagulopathy, mechanism of in contrast reaction, 169
Contact system activation, 173
  in contrast reactions, 176–177
Contrast allergy reaction, in children, 208
Contrast enema, for meconium ileus, 219–220. *See also* Barium enemas

Contrast-enhanced computed tomography
  advantage of, 67
  contraindications to, 66–67, 70–71
  contrast media available to, 74
  disadvantages of, 67
  dosage in, 74–75
  general indications for, 66–67
  management of contrast extravasation in, 80–82
  route of administration in, 75
  scan protocols for, 75–79
  specific indications for, 67–68
  techniques of contrast administration in, 68–70
  types of contrast used in, 71–74
Contrast media. *See also specific types of and agents*
  available, 74
  bolus technique of administration of, 68–69
  for bowel, 158
  causes of death with, 167
  causing damage to blood-brain barrier, 89
  chemical composition and toxicity of, 171–172
  choice of for hysterosalpingography, 139–142
  choice of for neuroangiography, 93
  in computed tomography, 66–82
  contraindications to
    in gastrointestinal studies, 201
    in intravenous administration of, 70–71
  CT portography injection of, 69
  in cystography, 63
  in cystourethrography in children, 216–217
  dose of, 28–29
    and reactions to, 166–167
    in computed tomography, 74–75
  drip infusion of, 68
  dynamic sequential bolus administration of, 69
  effect of injection of
    bolus phase of, 68
    equilibrium phase of, 68
    nonequilbrium phase of, 68
  effects of on renal blood flow, 29
  excretion of, 29
    peak time for, 29
  extravasation of, 73–74
    incidence of, 80
    mechanisms of toxicity of, 80
    patients at risk for, 82t
    presentation of, 80
    prevention of, 81–82
    treatment of, 80–81
    with venography, 104
  fistulographic, 205–206

in gastrointestinal tract of children, 212–216
half-life of, 53–54
in hepatic CT angiography, 69–70
high osmolar, 2
  vs. low osmolar, 54–56, 100–101, 215–216
high risk vs. low risk, 172
high vs. low osmolar in gastrointestinal tract evaluation in children, 215–216
hypertonic, 93
hypertonicity of, 112–113
for hysterosalpingography
  history of use of for, 135t
  unique features of, 140t
imaging techniques for detection with, 145–146
indications for use of in computed tomography in children, 211t
Infusaports and Portacaths for administration of, 75
injection rates and plasma concentration of, 10–12
injection technique
  and dose of in myelography, 114
  for neuroangiography, 91–92
intravasation of, 206
intravascular, properties and general effects of, 1–18
in intravenous urography
  chemistry of, 54–56
  concentration of, 56
  dose and volume of, 58–59
  physiology of, 52–54
  route of administration of, 59–60
  viscosity of, 56–58
iodine content, viscosity, and osmolality of, 74
ionic, 83–84
ionic vs. nonionic, 3, 71–73, 93
  reactions to, 19–20, 161–164
low-osmolality
  in digital subtraction angiography, 97
  indications for use of, 101t
low osmolar, 2, 35, 93
  economic issues in, 180–183
  legal considerations in, 183–186
  situations warranting use of in intravenous urography, 60t
for magnetic resonance imaging, 143–160
monomer vs. dimer, 3–4
for myelography, 107–109
  physiology of neurotoxicity in, 109–111
  symptomatology and origin of neurotoxicity with, 111–113
negative, 108, 202
in neuroangiography, 90–93

neurotoxicity, 89–90, 111–113
neutral, 202
nonionic, 84
  reasons for reduced reactions to with, 164–165
oil-soluble vs. water-soluble, 139–142
pathophysiology of, 84–90
pediatric, 207–221
physical factors of, 74t
positive, 202–204
  oil-based, 108–109
  water-soluble, 109
potential problems with, 93
rapid tissue clearance of, 145
rate of administration of, 60
reactions to
  and chemical composition and toxicity, 171–172
  factors in treatment of, 20–21
  incidence of, 161–164
  identification of, 22
  life-threatening, 99–100
  mechanisms of, 21–22, 161–179
  medical negligence and, 184–185
  with nonionic vs. ionic media, 19–20, 161–164
  possible explanations of, 167–170
  prevention of, 170
  prior major, 70
  range of, 166
  risk factors and prevention of, 100–101
  systemic, 23t, 171–177
  treatment of, 22–27, 170
  universal pretreatment options for, 21t
recommendations about concentration and volume of, 115t
with reduced toxicity, 34
renal enhancement by, 30
renal handling of, 28–30
route of administration of, 75
routine injections of, 75
stability of, 145
suggested use of oral, 201t
systemic reactions to, 23t, 171
  mechanisms of, 21–22
  pathogenesis of, 172–175
  patients at risk for, 175–177
  role of anxiety in, 175
  treatment of, 177–179
techniques of administration of, 68–70
therapeutic use of in children, 218–220
toxicity of, 145
transit of through pulmonary system, 11
treatment of acute reactions to, 19–27

Contrast media—*continued*
  tubular effects of, 29–30
  types of used in computed tomography, 71–74
  with urethrography, 65
  used for urography infusion technique, 57*t*
  used in digital subtraction angiography, 97
  used in urography rapid injection technique, 55*t*
  use of in central nervous system, 83–93
  use of with MRI in children, 211–212
  for venography, 102–103
    complications related to, 103–106
    high osmolar vs. low osmolar, 103
    indications for specific, 106
  vicarious excretion of, 29
  water-soluble, 1–2
    gastrointestinal, 200–201
    for gastrointestinal computed tomography, 202–203
Contrast medium-induced acute renal failure, 28, 31
  hypothetical mechanisms of, 31–32
  and new contrast agents, 34
  prevention of, 34–35
  radiologic findings in, 32
  risk factors for, 31*t*
  treatment of, 32–34
Contrast nephropathy, 41
  risk of with renal insufficiency, 60
Contrast neuropathy, patients at risk for, 59*t*
Convulsions, treatment of, 25–27
Cortical blindness
  contrast media-induced, 112
  following vertebral angiography, 89
Corticosteroids
  for anaphylactoid reaction, 24
  to prevent reactions, 170, 176, 177
  to treat contrast reactions, 24, 179
Cost-effectiveness analysis, 182–183
Crohn's disease
  detection of, 195
  diagnosis of in children, 214
Cutaneous fistulas, contrast agents for, 205–206
Cysto-Conray, 216
Cystography, 62
  for azotemia diagnosis, 44
  complications of, 64
  contrast media with, 63
  goals of, 62
  importance of, 50
  technique of, 62–63
Cystourethrography
  in children, 216–217
  voiding, 63–64, 216

Deep venous thrombosis (DVT)
  evaluation of, 102
  postphlebographic, 105
  postvenography, 106
Defibrillation, 25
Degenerative disc disease, detection of, 154–155
Dehydration, risk of with contrast media-induced nephrotoxicity, 30
Demyelinating diseases, detection of, 153
Demyelination, spinal, 156
Density, 6
Diagnostic tools, 36–44. *See also specific techniques*
Dialysis, 33
Diatrizoate, 216
Diazepam (Valium)
  in prevention reactions to contrast media, 170
  for seizure reaction, 27
Diffusion imaging, 145
Digital subtraction angiography (DSA), 1
  advantages of, 96
  contrast agents used for, 97
  indications for, 96
  intra-arterial, 91
  intravenous injection technique in, 91–92
  low osmolar contrast media in, 87
  for primary azotemia diagnosis, 45
  techniques and applications of, 96–97
Digital subtraction ateriography, 46
Dimers, 84
Dimer-X, 84
Dimethyl sulfoxide, for contrast extravasation, 80
Dionosil, 217
Diphenhydramine (Benadryl)
  for anaphylactoid reaction, 24
  for hives, 23
Disc herniation, 154
Disodium edetate, reduced ionic calcium with, 17–18
Distal radioulnar joint injection, 126–127
Distribution volume, 8
DMSO, 4
Dopamine
  for hypotension, 24
  for treatment of reactions, 170
Doppler ultrasonography, 45
Dotarem, 149
Double-balloon catheter, for female urethra, 65
Double-contrast barium enemas, 189–190, 195–197
Double-J catheter, 42–43

INDEX 229

Drip infusion technique, 68
DTPA, use of with MRI in children, 212
Dural arteriovenous malformations, 154
Dynamic imaging techniques, 146
  in detection of cerebral infarction, 153
Dynamic sequential bolus computed tomography, 69
  for liver evaluation, 77–78
  reduced extravasation with, 73–74
  superiority of, 70
Dysreflexia, autonomic, 64

Ectopic pregnancy, diagnosis of, 131
Effervescent agents, 189–191
Elbow
  arthrographic technique for, 126
  lateral projection of, 126f
Electrolyte monitoring, in CM-ARF treatment, 33–34
Empyema, diagnosis of, 67
Endometrial carcinoma, detection of, 159
Endoscopic retrograde cholangiopancreatography (ERCP), 218
Endothelial damage, 17
Enteroclysis, 193–194
  double-contrast study, 194
Enzymuria, abnormal, 31
Ependymomas, 155
Epinephrine (Adrenalin)
  for anaphylactoid reaction, 24
  for bronchospastic reaction, 24
  to treat contrast reactions, 177–178
  for unconscious, unresponsive, pulseless reaction, 25
Esophageal varices, 192
Esophagotracheal fistulas, hyperosmolar ionic contrast media with, 201
Esophagus, barium sulfate studies of, 191–192
Ethylene glycol, 4
Excretory urography
  for azotemia diagnosis, 42
  in children, 209–210
  nephrographic patterns and time-density curves with, 32, 33f
  in secondary azotemia diagnosis, 48–49
Extracellular space, contrast media distribution through, 8
Extracellular tracers, 8
Extradural space, contrast enhancement of, 156
Extraperitoneal perforation, 198–199

Factor XII, activation of, 174, 175
Fat suppression imaging techniques, 157
Femoral artery catheterization, 69
Fever, mechanism of, 169
Fibrinolysin system activation, 173
Fistulas
  causes of, 205
  gastrointestinal, 215
Fistulographic contrast agents, 205–206
Fluid management, for CM-ARF treatment, 32
Fluoroscopy, biplane, 205
Foley catheter, transurethral placement of, 62
Foreign body detection, 192
Foreign body granulomas, with hysterosalpingography, 139

Gadodiamide (Gd-DTPA-BMA), 148–149
Gadolinium diethylenetriamine pentaacetic acid (Gd-DTPA), 147–150
  use of with MRI in children, 212
Gadopentetate dimeglumine
  altered tissue relaxation with, 143
  complications with, 160
  crossing of placenta by, 159
  in detection of acute myocardial infarction, 157
  elimination of, 159–160
  extracellular distribution of, 144–145
  intravenous, patient preparation for, 150–151
  in kidney scanning, 158
  for liver and spleen scanning, 157–158
  transient hemolysis with, 160
Gadoteridol (Gd-HP-DO3A), 148
Gallbladder, CT protocol for, 78
Gastrografin, 201
  for gastrointestinal computed tomography, 203
  for meconium ileus, 219
Gastrointestinal tract
  barium sulfate use in, 187–199
  contrast agents
    in children, 212–216
    in computed tomography of, 202–204
    for detection of perforation of, 215
    water-soluble, 200–201
Glomerular filtration
  contrast media excretion by, 52
  of intravascular contrast media, 10f
  of X-ray contrast media, 8–9
Glomerular filtration rate, 28

Glomerular filtration rate—*continued*
  reduction in, 30
Glucagon
  in abdominal digital subtraction angiography, 97
  hypotonia with, 194
Gradient-echo imaging, 145, 146

H_2 blockers
  to prevent reactions, 177
  for urticaria, 23
Head
  arteriovenous abnormalities of, 154
  cerebral infarction of, 153–154
  demyelinating diseases of, 153
  indications for magnetic resonance imaging of, 151–154
  infection of, 153
  neoplasia of, 151–153
  neuroangiography for trauma to, 90
Heavy-metal scavengers, 4
Hemangioma, differentiation of, 158
Hemodynamics
  contrast media effects on, 29
  effects of intravascular contrast media on, 12–13
  reduced changes in with nonionic agents, 164
Hemolytic anemia, as contraindication for MRI, 160
Hepatic artery injection computed tomography, 69
Hepatic computed tomographic angiography, 69–70
Hepatic computed tomography, routine protocol for, 77–78
Hepatobiliary contrast agents, 149
Hepatocellular carcinoma, differentiation of, 158
Herniated lumbar disc, 115
Hexabrix, 84
  reactions to, 100–101
Hickman catheter, 75
High osmolar contrast media (HOCM), 2
  chemistry of, 54–56
  in gastrointestinal tract evaluations in children, 215
  physiologic derangements with in children, 210
High-risk patients, need to identify, 184
Hip
  anteroposterior line drawing of, 123*f*
  arthrographic technique for, 122–124
  films of, 124
  total joint implants of, 124

Histamine blockers, to prevent reactions, 170
Histamine release
  with contrast media, 165, 173
  in contrast reactions, 21, 177
History, unknown or unobtainable, 106
Hives, treatment of, 23
Hospital liability, 183–184
HP-DP3A, chemical structure for, 149*f*
Hyaluronidase, for contrast extravasation, 81
Hydrocortisone, for anaphylactoid reaction, 24
Hydronephrosis
  causes of false negative ultrasonographic diagnoses for, 37*t*
  detection of, 36
  ultrasonographic diagnosis with, 36–37
Hypaque
  in cystourethrography, 216
  for gastrointestinal computed tomography, 203
  for meconium ileus, 219
Hypaque 20, 97
Hypaque-90 media, 52
Hyperkalemia
  contrast media with, 71
  dialysis for, 33
Hyperosmolality, and risk of contrast reactions, 167
Hypersensitivity reactions, to barium sulfate, 197–198
Hypertensive intravenous urogram, 61
Hyperventilation, 138
Hypocalcemia, with intravascular contrast media, 17–18
Hypokalemia, 71
Hypotension
  with contrast media reaction, 99, 168
  identification of, 22
  neuroangiography with, 92–93
  treatment of, 24
Hypothalamus, body temperature control by, 169
Hypovolemia, hypertonic agents with, 201
Hysterosalpingography
  cannulas used in, 133, 134*f*
  choice of contrast agent for, 139–142
  complications with, 135–139
  contraindications for, 132
  contrast media used in, 135*t*
  films taken in, 134–135
  indications for, 131, 132*t*
  limitations of, 131

INDEX   **231**

pregnancy rates after, 139
technique of, 132–135

Imaging. *See also specific techniques*
for azotemia patient, 36–40
in primary azotemia diagnosis, 44–47
in secondary azotemia diagnosis, 47–49
Immunoglobulin E, and contrast media reactions, 173
Indwelling intravenous lines, 81–82
Infertility
diagnosis of, 131
hysterosalpingography for, 141*f*
Infusaports, 75
Infusion technique
administration route in, 59
for postoperative patients, 61
situations warranting use of, 58–59
Injection techniques, for intravenous contrast media, 10–13
Intra-abdominal fluid collections, 67
Intra-arterial injection
physiologic effects of, 14–16
rate and blood-contrast media mixing in, 13–14
Intracerebral hemorrhage, neuroangiography for, 90
Intracoronary injections, physiologic effects of, 15–16
Intracranial blood flow, reduction of, 87
Intrathecal air injection, 83
Intrauterine devices, location of, 131
Intravasation, with hysterosalpingography, 137, 138*f*
Intravascular contrast media
acute lethal dose of, 8
annual volume of, 1
clearance of, 8–10
compared to, 1
complement activation of, 18
density of, 6
distribution volume of, 8
endothelial damage and thrombosis with, 17
examinations requiring, 1
general considerations in use of, 1–2
hypocalcemia with, 17–18
injection rates and plasma concentration of, 10–12
injection techniques for, 10–13
intra-arterial injections of
physiologic effects of, 14–16
rate of and blood-contract media mixing, 13–14
molecules and additives of, 3–4
osmolality of, 7
partition coefficient of, 7

penetration of into central nervous system, 16
pharmacokinetics of, 8–10
physical and chemical properties of solutions of, 4–8
physical characteristics of, 2
physical principles of related to imaging, 2–3
physiologic actions of, 12–13
pulmonary edema with, 18
therapeutic agents used in, 1
urinary iodine concentrations with, 16–17
viscosity of, 6–7
Intravascular enhancement, phases of, 76–77
Intravenous contrast media
extracellular distribution of, 147–149
hepatobiliary, 149
particulate, 149
Intravenous urography, decreased number of, 50
Intravoxel incoherent motion, 145
Intussusception, reduction of in children, 220–221
Iodinated contrast agents, intravascular
accidents with, 208
anaphylactic reaction to, 160
idiosyncratic reaction to in children, 208–209
for pediatric patients, 207–209
toxicity of in children, 208
Iodinated contrast media, intravenous, osmolality of, 209*t*
Iodinated water-soluble agents, 1
Iodine
additive of, 4
aortic plasma concentration of, 11, 12
arteriovenous differences in, 76
blood concentration of vs. time after bolus injection, 54*f*
concentration of and osmolality, 99*t*
maximum urinary concentration of, 58
peak urine concentrations of, 29
ratio of to dissolved particles, 3–4
urinary concentrations of, 16–17
Ionic contrast media, 3, 71
binding of ionic calcium by, 13
causing damage to blood-brain barrier, 89
decreased blood coagulation with, 86
development of, 83–84
incidence of reactions to, 161
indications for in gastrointestinal tract, 200–201
injuries caused by, 73–74

Ionic contrast media—*continued*
  intracoronary injection of, 15–16
  reactions to, 19–20
  safety of compared with nonionic media, 72–73
  viscosity of, 56
Iopamidol, ionized calcium in blood with, 165
Iothalamate, 216
Ioxaglic acid, ionized calcium in blood with, 165
Isoproterenol, for contrast extravasation, 80

Joint effusions, 120
Joints
  aspiration of fluid from, 118–119
  sterile preparation of, 118

Kallikrein, stimulation of production of, 18
Kidney(s)
  contrast media excretion by, 29
  MRI scanning of, 158
  routine computed tomography protocol for, 79
Knee, arthrographic technique for, 119–120
K-shell-binding energy (K edge), 2
KUB plain film, 51

Lactation, as contraindication for MRI, 159
Lasix intravenous urogram, 61–62
Left ventricular pressure, intracoronary injection of ionic contrast media and, 15
Legal issues, 183–186
Liability, for negligence, 183–185
Limb, compromised, 106
Limbic system, function of, 168
Linear attenuation coefficient, 2
Lipid solubility, 7
Lipiodol, introduction of, 107
Liquid effervescent agents, 190–191
Liver
  determining resectability of tumors of, 69–70
  metastasis of, 157, 158
    hypervascular, 67
    local iodine concentration in, 69
  MRI scanning of, 157–158
  peak tissue enhancement of, 76
  routine CT protocol for, 77–78
Low osmolar contrast media (LOCM), 2. *See also* Nonionic contrast media
  benefit analysis of, 181–182
  chemistry of, 54–56
  cost analysis of, 180–181
  cost-effectiveness analysis of, 182–183
  in cystography, 63
  development of, 180
  economic issues for, 180–183
  in gastrointestinal tract evaluations in children, 215–216
  increased cost of compared to high osmolar contrast media, 180
  legal considerations in, 183–186
  percentage of patients receiving, 181
  safety and efficacy of, 184
  selection for in children, 210*t*
  situations warranting use of in intravenous urography, 60*t*
Lymphatic intravasation, with hysterosalpingography, 137

Magnetic iron oxide particles, 150
Magnetic resonance contrast agents
  complications with, 160
  contraindications to, 159–160
  design of, 143–145
  imaging techniques for detection with, 145–146
  indications for
    in body, 156–159
    in head, 151–154
    in spine, 154–156
  intravenous, 147–149
  mechanism of enhancement of, 143–144
  oral, 149–150
  patient preparation for use of, 150–151
  tissue specificity of, 144–145
Magnetic resonance imaging (MRI), 1
  for azotemia diagnosis, 43
    primary, 47
  bulk magnetization in, 143
  in children, 211–212
  increasing availability of, 108
  limitation of, 50
  vs. myelography, 115–116
  with pheochromocytoma, 70
  relaxivity-based contrast enhancement in, 144
  signal intensity in, 143
  signal loss in, 144
  susceptibility of substance in, 143–144
  tissue specificity in, 144–145
Magnevist, 147–148
Meconium ileus
  benefits of contrast enema in, 219–220

complications of contrast media in, 219
indications for contrast media in, 218
pathologic nature of, 218–219
therapeutic contrast enemas for, 219
Mediastinal structure, defining of with computed tomography, 67–68
Medical malpractice, 183
with contrast media reactions, 184
Medrol (methylprednisolone), to prevent reactions, 176
Meglumine (methylglucamine), 83–84
Meglumine salts, viscosity of, 2
Meningeal disease, detection of, 152
Meningiomas, detection of, 155–156
Meningitis, 155
Mesenteric artery
contrast media injection of, 95
visualization of, 94
Metaiodobenzylguanimidine (MIBG) radionuclide study, 70
Metal chelates, 147–149
positive enhancement, 149–150
Metal ion retention, chronic toxicity due to, 159–160
Methylcellulose solution, 190
in antegrade small bowel examination, 193–194
Methylprednisolone (Solu-Medrol), for anaphylactoid reaction, 24, 176
Metrizamide, in bronchography in children, 217–218
Midcarpal compartment injection, 126
Migraine headache, neuroangiography with, 92
Mn-DPDP, 149
Molal concentration, 6
Molar concentration, 6
Monomeric agents, osmolality of, 6
Multiple myeloma, contrast media with, 71
Musculoskeletal system, MRI scanning of, 159
Myelography
contrast media for, 107–109
injection and dose in, 114
symptomatology and origin of neurotoxicity in, 111–113
indications for, 115–116
introduction of, 107
measures following, 114–115
physiology of neurotoxicity in, 109–111
preparation and precautions for, 113–114
Myocardial infarction, acute, detection of, 157

Nasogastric tube gastrointestinal studies, 189–190
Nausea
mechanism of, 169
treatment of, 22–23
Neck trauma, neuroangiography for, 90
Necrosis, 80
Negligence, 183–185
Neoplasia
of head
detection of extra-axial, 151–152
detection of intra-axial, 152–153
musculoskeletal, 159
spinal, 155–156
Nephrogram, 51
Nephrostomy, percutaneous, 46
Nephrotomograms, 50
Nephrotoxicity, contrast medium-induced, 28–35
Neuroangiography
choice of contrast medium for, 93
contraindications to, 92–93
contrast media for, 83–84
injection technique for, 91–92
indications for, 90
preparation of patient for, 91
Neurofibromas, detection of, 155–156
Neurofibromatosis, 156
Neurotoxicity
and central nervous system physiology, 110–111
in children, 208
contrast media, 89–90
symptomatology and origin of, 111–113
Noninvasive modalities, improvement in, 101
Nonionic contrast media, 3
ACR guidelines for use of, 71–72
current criteria for use of, 20t
development of, 84
improved patient tolerance of, 72
incidence of reactions to, 161
indications for in gastrointestinal tract, 201
to prevent reactions, 170
reactions to, 19–20
compared with ionic media, 19–20, 161–164
reasons for reduction of, 164–165
reduced hemodynamic changes with, 164
reduced risk of extravasation injury with, 82
safety of compared with ionic media, 72–73
viscosity of, 56–58
Nonopaque cation meglumine, filtration of and tubule osmolality, 16–17

Nuclear imaging, for azotemia diagnosis, 41–42
  primary, 45
  secondary, 49

Obese patients
  contrast media distribution in, 59
  infusion technique for, 58
Oil-soluble contrast media, for hysterosalpingography, 139, 141f
Oliguria, in contrast medium-induced acute renal failure, 31
Omnipaque
  injection and dose of, 114
  neurotoxicity of, 112
Oral contrast media, 149–150
Oral magnetic particles, 150
Orthoiodohippurate, 41
Osmolality, 6, 7
Osmolar toxicity, in children, 208
Osmotic diuresis, 30
  intravascular, 17
Osmotic nephrosis, 31
Osteomyelitis, 155
Osteonecrosis, detection of, 159
Oxygen
  for anaphylactoid reaction, 24
  for bronchospastic reaction, 23
  for hypotension, 24

Pain
  with hysterosalpingography, 135–136
  postarthrography, 117–118
  with venography, 104
Pancreas
  CT protocol for, 78
  peak tissue enhancement of, 76
Pantopaque, 108
  density and properties of, 109
  introduction of, 107
  low toxicity of, 108
Paramagnetic contrast agents, relaxivity-based enhancement by, 144
Paramagnetic metal ion chelates, extracellular distribution of, 144–145
Parasympathetic system, activation of, 169
Particulate agents, 149
  spin dephasing with, 144
Partition coefficient, 7
Pediatric contrast agents
  angiocardiography and angiography, 210
  biliary, 218
  bronchography, 217–218
  computed tomography, 210–211
  cystourethrography, 216–217
  excretory urography, 209–210
  gastrointestinal tract, 212–216
  general considerations in, 207
  intravascular iodinated, 207–209
  intravenous, for magnetic resonance imaging, 211–212
  intussusception of, 220–221
  need for individualized dosage of, 207
  therapeutic use of, 218–220
Pelvis
  infection of with hysterosalpingography, 136–137
  MRI scanning of, 159
  routine CT protocol for, 78–79
Pelvocalyceal collecting system
  distension and opacification of, 50
  modalities of visualization of, 50
Peripheral angiography, 95–96
Peripheral vacular disease, compromised limb with, 106
Peripheral venography
  complications of, 80
  contrast media for, 102–103
  contrast-related complications with, 103–106
  general considerations for, 102
  indications for specific contrast agents in, 106
Peristalsis, increased with contrast agents, 200–201
Peroral pneumocolon, 195
Pheochromocytoma, contrast media with, 70–71
Pigtail catheter, 94
Pituitary microadenomas, identification of, 153
Plain film
  abdominal, 36
  evaluation for intravenous urography, 51
  in urinary tract evaluation, 50
Plasma contact system, activation of, 174–175
Plasma viscosity, 85
  and contrast media, 85–86
Plasma volume, increased, 86–87
Platelets, activation of with contrast media, 18
Pleural space disease, evaluation of, 68
Plexiform neurofibromas, 156
Pneumatic reduction, of intussusception in children, 220
Pneumomyelography, introduction of, 107
Pneumonia, with barium aspiration, 197
Portacaths, 75

Portography, computed tomographic, 69, 78
Postoperative patients, imaging in, 61
Postphlebography syndrome, 105
Potassium, abnormal serum levels of, 71
Power injector, and contrast extravasation risk, 73
Pregnancy
 as contraindication for hysterosalpingography, 132
 as contraindication for MRI, 159
 radiation exposure during early stage of, 138–139
 rates of after hysterosalpingography, 139
Prekallikrein
 activation of, 174
 transformation of in contrast reaction pathway, 21
ProHance, 148
Propranolol, for contrast extravasation, 80
Prosthetic joints, sterile preparation of, 118
Pseudoaneurysm, neuroangiography for, 90
Pulmonary arterial pressure (PAP)
 elevation of with contrast media in pulmonary vasculature, 12
 increase in with contrast media injection, 18
Pulmonary edema
 in children, 208
 with intravascular contrast media, 18
 mechanism of in reaction to contrast media, 169
Pulmonary embolization
 mortality rate with, 103–104
 with venography, 103
Pulmonary hemorrhage, 208
Pulmonary toxicity, 208
Pyelography
 for azotemia diagnosis, 42–43, 46, 49
 in intravenous urography, 51–52
 quality of, 51

Radiation exposure, with hysterosalpingography, 138–139
Radiocarpal compartment injection, 126
Radiographic contrast media, principles of relating to imaging, 2–3
Radiography, supine, 44
 for azotemia primary diagnosis, 46
Ranitidine (Zantac)
 for anaphylactoid reaction, 24
 for urticaria, 23

Rapid drip infusion, 59
Reactions, types of, 171. *See also* Contrast media, reactions to
Reasonable patient standard, 185
Red blood cells
 concentration of, 85
 crenation and aggregation of, 17
 deformability of, 85
 deformed, 86
 dessication of with contrast media injection, 12–13, 14*f*
 effect of hypertonic contrast media on, 86
 intact, deformable, 86
 internal viscosity of, 85–86
 ionic contrast media effects on, 86–87
 reduced deformation and aggregation of with low osmolar agents, 164
Relaxivity-based imaging, 146
Relaxivity contrast agents, 146
Renal angiography, 94–95
Renal arteries, visualization of, 94
Renal biopsy
 percutaneous, 43–44
 in primary azotemia diagnosis, 44–45
 in secondary azotemia diagnosis, 49
Renal blood flow, contrast media effects on, 29
Renal clearance
 half-time for, 9
 of X-ray contrast media, 8–10
Renal failure
 acute, 28–35, 38–39
 chronic, 38–39
 as contraindication for MRI, 159–160
 contrast-related, 28–35, 70, 98–99
 excretory urography for, 42
 intrarenal causes of, 38–39
Renal insufficiency
 imaging in, 60–61
 with venography, 106
Renal parenchymal disease, biopsy for, 43–45
Renal stone disease, infusion technique for, 59
Renal system
 function of
  compromised, infusion technique for, 59
  MRI assessment of, 158
  MRI detection of masses in, 158
  routine CT protocol for, 79
Renal vein thrombosis (RVT)
 as cause of renal failure, 43
 renal venography with, 45–46
 unilateral, 43
Renal venography, 45–46

Renografin
  in cystourethrography, 216
  for hysterosalpingography, 142
Renography
  in azotemia diagnosis, 41–42
  for primary azotemia diagnosis, 45
Reticuloendothelial system (RES), scavenging of, 149
Retrograde pyelography, 42–43
  for azotemia primary diagnosis, 46
Retrograde small bowel examination, 194–195
Retrograde urethrography, 65

Salutar, 148–149
Scanner software improvements, 75–76
Schwannomas, detection of, 155–156
Seizures
  with Amipaque myelography, 113–114
  in children, 208
  identification of, 22
  after intravenous contrast media administration, 16
  treatment of, 25–27
Shock, in reactions to contrast media, 168
Shoulder
  anteroposterior view of, 121f
  arthrographic technique for, 120–122
  films of, 122
  intra-articular needle placement in, 120–121
  superficial needle placement in, 122
Sickle cell anemia, contrast media with, 71
Signal loss, with diffusion, 145
Single-contrast barium enema, 195–196
Single-contrast barium stomach examination, 192
Sinografin, 142
Skin ulceration, with contrast extravasation, 80
Small bowel
  barium studies of, 193–195
  method of opacifying, 204
  obstruction of, 193
Sodium citrate, reduced ionic calcium with, 17–18
Sodium salts, viscosity of, 2
Spinal angiography, damage caused by, 90
Spinal cord
  contrast media injury of, 90
  contusion of, 154
  infarcts of, 156
  metastasis of, 155

Spinal tap headache, reducing risk of, 114–115
Spin density assessment, 145
Spine
  congenital lesions of, 156
  degenerative disc disease of, 154–155
  demyelination of, 156
  indications for magnetic resonance imaging of, 154–156
  inflammatory lesions of, 155
  neoplasia of
    extradural, 156
    intradural-extramedullary, 155–156
    intramedullary, 155
  vascular disease of, 156
Spin-echo imaging techniques, 146
Spleen, MRI scanning of, 157–158
Staphylococcus epidermidis infection, postarthrography, 118
Steroids
  for contrast extravasation, 80
  to treat contrast reactions, 179
Stomach
  barium sulfate studies of, 192
  method of opacifying, 204
Stress, and risk of contrast reaction, 21
Strontium bromide, 83
Subarachnoid hemorrhage, neuroangiography for, 90
Subclavian steal phenomenon, 90
Surgical drainage, for contrast extravasation, 81
Susceptibility contrast enhancement, 145–146
Systemic reactions, 171
  mediators and activation systems in, 172–173
  pathogenesis of, 172–175

T1-weighted techniques, 146, 147f
  uterine, 159
T2-weighted imaging, 158
Tachycardia, with contrast media reaction, 99
Technetium-99m-DMSA, 42
Technetium-99m-DTPA, in renography, 41–42
Technetium-99m-GHA, 42
Temporomandibular joint
  arthrographic technique for, 128–130
  composition of, 128
  excessively posterior needle position in, 130
  films of, 130
  intra-articular needle position in, 129

transcranial lateral view of, 129f
Thin collimation
  for bile duct and gallbladder evaluation, 78
  for pancreatic evaluation, 78
Thoracic aortic dissection, CT evaluation of, 79
Thrombosis, with intravascular contrast media, 17
Tibiotalar joint, 124
Tissue enhancement, peak of, 76
Tissue specificity, of magnetic resonance contrast agents, 144–145
Transitory ischemic attacks, neuroangiography for, 90
Transverse myelitis, 155
Triiodinated benzene rings, 5f
Tube heat capacity, 75
Tubular nephrogram, dense, 53
Tubular system
  contrast media effect on, 29–30
  opacification of, 51
  sodium and water reabsorption in, 30
Turbo-FLASH technique, 146
  in detection of cerebral infarction, 153

Ultrasonography
  for azotemia diagnosis, 36–40
    primary, 44, 46, 47
    secondary, 47–48
  false positive diagnoses with, 37
  for renal failure diagnosis, 37–40
Unconsciousness, treatment of, 25
Uremia, dialysis for, 33
Ureteral brush biopsy, 46
Ureteroscopy, 46
Ureters
  compression of, 51
  distension and opacification of, 50
  visualization of, 52
Urethral diverticula, 65
Urethral injury, cystogram of, 652–653
Urethrography
  complications with, 65
  contrast media with, 65
  goals of, 64
  importance of, 50
  technique of, 64–65
Urinary bladder
  adequate distension of, 52
  distension and opacification of, 50
  infections of with cystography, 64
  intraluminal pressures of, 63–64
  postvoid film of, 52
  tears in, 44, 63
  visualization of, 52
Urinary iodine concentrations, 16–17

Urinary tract, plain film evaluation of, 50
Urine flow rates, with various contrast media, 29
Urography
  for azotemia diagnosis, 42
  infusion technique
    commonly used intravenous contrast media for, 57t
    situations that may warrant use of, 58t, 59
  intravenous
    contrast media in, 52–60
    goals of, 50
    hypertensive, 61
    Lasix, 61–62
    preparation for, 51
    in special conditions, 60–62
    technique of, 51–52
  rapid injection technique, commonly used intravenous contrast media for, 55t
Urticaria
  with hysterosalpingography, 137
  mechanism of in reaction to contrast media, 169
  treatment of, 23
Uterus
  diagnosis of cancer of, 131
  differentiation of zonal anatomy of, 159
  evaluation of anomalies of, 131

Vacuolization, with hypertonic contrast agents, 31
Vagal reaction
  identification of, 22
  treatment of, 24–25
Vaginal bleeding, 132
  with hysterosalpingography, 137
Vascular bed, contrast media pathophysiology in, 87–89
Vascular blush, 51
Vascular imaging, improvement in noninvasive modalities for, 101
Vasoactive prostaglandins, production of, 176–177
Vasoactive substances, release of, 165
Vasodilatation
  with hyperosmotic contrast media, 87
  of intracranial vessels, 87–88
Vasopressors, for hypotension, 24
Vasospasm, effect of on intracranial vessels, 87–88
Vasovagal reactions, 137
Venography
  for azotemia diagnosis, 43
  complications of, 80
  in deep venous thrombosis diagnosis, 102

Venography—*continued*
  pain with, 104
  peripheral, 102–106
  for primary azotemia diagnosis, 45–46
Venous angiomas, 154
Venous intravasation, with hysterosalpingography, 137, 138*f*
Vesicoureteral reflex, 64
  assessment of in children, 216
  documentation of, 44
Vicarious excretion, 8
Visceral angiography, 94–95
Viscosity, 6–7
  definition of, 2
Voiding cystourethrography (VCU), in children, 216
Volume expansion, in CM-ARF treatment, 32
Volume overload, dialysis for, 33
Vomiting
  mechanism of, 169
  treatment of, 22–23

Water-soluble contrast media
  in bronchography in children, 217–218
  in gastrointestinal studies, 200–201
  in children, 215
  computed tomography, 202–204
  high osmolality and viscosity of, 1–2
  for hysterosalpingography, 139, 141*f*, 142
  for meconium ileus, 219–220
  for myelography, 107, 109
  for reduction of intussusception in children, 220
  uses of in gut, 219
Wilson's disease, as contraindication for MRI, 160
Wrist
  arthrographic technique for, 126–128
  films of, 128
  posteroanterior line drawing of, 127*f*
  three-compartment arthrogram of, 126–127

X-ray contrast media
  clearance of, 8–10
  distribution volume of, 8
  triiodinate benzene ring of, 5*f*
X-ray detector, 2–3
X-ray radiation, 2